Conference Proceedings of the Society for Experimental Mechanics Series

Series Editor

Kristin B. Zimmerman, Ph.D
Society for Experimental Mechanics
Bethel, CT, USA

More information about this series at http://www.springer.com/series/8922

Srinivasan Arjun Tekalur • Pablo Zavattieri • Chad S. Korach
Editors

Mechanics of Biological Systems and Materials, Volume 6

Proceedings of the 2015 Annual Conference on Experimental and Applied Mechanics

Springer

Editors
Srinivasan Arjun Tekalur
Michigan State University
East Lansing, MI, USA

Pablo Zavattieri
Purdue University
West Lafayette, IN, USA

Chad S. Korach
Department of Mechanical Engineering
State University of New York at Stony Brook
Stony Brook, NY, USA

ISSN 2191-5644 ISSN 2191-5652 (electronic)
Conference Proceedings of the Society for Experimental Mechanics Series
ISBN 978-3-319-21454-2 ISBN 978-3-319-21455-9 (eBook)
DOI 10.1007/978-3-319-21455-9

Library of Congress Control Number: 2015945112

Springer Cham Heidelberg New York Dordrecht London

Printed on acid-free paper

Springer International Publishing AG Switzerland is part of Springer Science+Business Media (www.springer.com)

Preface

Mechanics of Biological Systems and Materials represents one of nine volumes of technical papers presented at the SEM 2015 SEM Annual Conference and Exposition on Experimental and Applied Mechanics organized by the Society for Experimental Mechanics and held in Costa Mesa, CA, June 8–11, 2015. The complete Proceedings also include volumes on Dynamic Behavior of Materials; Challenges In Mechanics of Time-Dependent Materials; Advancement of Optical Methods in Experimental Mechanics; Experimental and Applied Mechanics; MEMS and Nanotechnology; Mechanics of Composite and Multifunctional Materials; Fracture, Fatigue, Failure and Damage Evolution; and Residual Stress, Thermomechanics and Infrared Imaging, Hybrid Techniques and Inverse Problems. Each collection presents early findings from experimental and computational investigations on an important area within Experimental Mechanics, the Mechanics of Biological Systems and Materials being one of these areas. This volume summarizes the exchange of ideas and information among scientists and engineers involved in the research and analysis of how mechanical loads interact with the structure, properties, and function of living organisms and their tissues. The scope includes experimental, imaging, numerical, and mathematical techniques and tools spanning various length and time scales. Establishing this symposium at the Annual Meeting of the Society for Experimental Mechanics provides a venue where state-of-the-art experimental methods can be leveraged in the study of biomechanics. A major goal of the symposium was for participants to collaborate in the asking of fundamental questions and the development of new techniques to address bio-inspired problems in society, human health, and the natural world. The organizers would like to thank all the speakers and staff at SEM for enabling a successful program.

East Lansing, MI, USA
West Lafayette, IN, USA
Stony Brook, NY, USA

Srinivasan Arjun Tekalur
Pablo Zavattieri
Chad S. Korach

Contents

Chapter 1
Damage in Brain Tissue Due to Single Bubble Cavitation Shock

Ghatu Subhash, Saranya Canchi, Yu Hong, Malisa Sarntinoranont, and Michael A. King

Abstract A novel experimental technique was developed to visualize and control single bubble cavitation initiation, growth, and its collapse. The influence of this process on a nearby tissue surrogate was investigated and then extended to a rat brain tissue. The technique utilized a modified polymer Hopkinson pressure bar system which transmits a simulated blast pressure wave with over and under pressure components to a fluid-filled test chamber implanted with a seed gas bubble. Growth and collapse of this bubble was visually recorded with a high speed camera. Using Raleigh-Plessat equation, bubble collapse pressures 29–125 times that of peak blast overpressure are predicted to be the source of localized shock waves. Finally, the value of this experimental platform to investigate the single bubble cavitation-induced fluid jet impact on a rat brain tissue and the associated damage evolution is illustrated.

Keywords Cavitation • Traumatic brain injury • Agarose hydrogel • Shock wave • Brain tissue

1.1 Introduction

Blast-induced traumatic brain injury (bTBI), a 'signature injury' of modern day war fare, has affected several thousands of soldiers returning from the recent wars [1]. Symptoms of these patients often range from mood swings to memory loss. Numerous possible mechanisms have been proposed including injury due to rapid accelerating—decelerating forces in the brain, multiple wave reflections with in the brain, impedance mismatch between heterogeneous tissue types, and cavitation [2]. Here we concern ourselves with cavitation caused by blast waves consisting of large overpressure (compressive) and smaller underpressure (tensile) components. The blast wave impinges on the head, leading to a portion of the wave being transmitted into cerebrospinal fluid (CSF) and brain tissue, both of which carry dissolved gasses. Within the short time period of blast wave transmission, there is potential for cavitation bubbles to form (during under-pressure phase), grow, and collapse within fluid spaces inside the head. When these bubbles collapse, considerable energy is released either in the form of local shock waves or generation of a high-speed fluid jet which may cause additional damage as it impinges on tissue. Such cavitation damage of material surfaces is well established since the original work of Rayleigh who first studied pitting of ship propeller blades or water turbine blades [3, 4] which is commonly referred to as 'cavitation erosion'. Thus, collapsing cavitation bubbles in the brain may cause local damage to nearby tissues or blood vessels.

A fundamental problem in the study of cavitation is that a testing platform is not readily available to control bubble growth and collapse. In this study, a novel experimental system capable of investigating single bubble cavitation was developed. The system was first utilized to quantify the dynamic behavior of a tissue surrogate from single bubble collapse. In the second phase the system is used to investigate damage in a biological tissue.

G. Subhash (✉) • S. Canchi • Y. Hong • M. Sarntinoranont
Department of Mechanical and Aerospace Engineering, University of Florida, Gainesville, FL 32611, USA
e-mail: subhash@ufl.edu

M.A. King
Department of Pharmacology and Therapeutics, University of Florida, Gainesville, FL 32611, USA

S.A. Tekalur et al. (eds.), *Mechanics of Biological Systems and Materials, Volume 6*,
Conference Proceedings of the Society for Experimental Mechanics Series, DOI 10.1007/978-3-319-21455-9_1

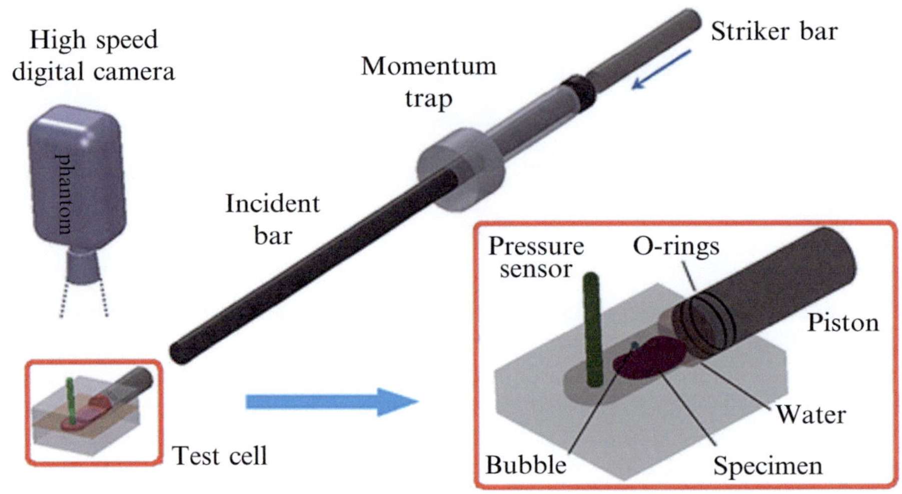

Fig. 1.1 Schematic of the momentum-trap PSHPB experimental set-up for cavitation testing of tissues

1.2 Experimental Method

A polymer split Hopkinson pressure bar (PSHPB) with momentum trapped incident bar [5] is used to generate lower magnitude pressure waves in an acrylic test cell containing fluid and the biological tissue. A schematic of the test system is shown in Fig. 1.1. A striker bar was launched from a gas gun towards the incident bar. This impact sends a stress wave through the incident bar, and upon reaching the piston, the stress wave accelerated the piston rapidly. After the initial compression, a momentum trap pulls back the incident bar in order to avoid multiple loading of the test cell. The rapid movement of the piston loads the fluid in the test chamber and generated a pressure profile similar to a blast wave. The transparent acrylic test cell consists of a chamber filled with artificial cerebral spinal fluid (aCSF) and a tissue slices. The cylindrical chamber was machined for a tight fit with the piston rod. An O-rings mounted on the piston provided a leak-proof seal. The pressure inside the chamber was measured utilizing a high rate pressure sensor with a sample rate of 100 kHz. In addition, a high speed digital camera was aligned above the test cell to view cavitation events at a frame rate of 80,000 frames/s.

After filling the chamber with aCSF and a tissue, the assembled test cell was submerged in water and was degassed to prevent the formation of multiple bubbles. A single seed bubble with a diameter of less than 50 μm was placed on inside wall of the submerged test cell.

Agarose-based hydrogel slices were prepared in a Petri dish and cut into suitable sizes. To allow tracking of deformation, hydrogel slices were speckled with black ink. These speckles move with the tissue slice upon loading and by capturing the time resolved images of the deformed specimen and using digital image correlation (DIC) one can measure full-field displacement and strain. The seed bubble was placed beneath the hydrogel close to the center of the slice so as to visualize the local shock wave response during bubble growth and collapse.

1.3 Results

Figure 1.2 shows temporal evolution of a typical bubble growth and associated pressure profile when the stress wave is imparted to the test cell. The seed bubble initially compresses during the compressive overpressure phase (a–c), grew during the tensile underpressure phase (d–h) and collapsed (n). The time duration from the bubble started to grow until collapse ranged from 270 to 350 μs. There was also secondary growth (o) followed by secondary collapse of the bubble (p–r). The bubble remained approximately spherical throughout the test until collapse.

Bubble growth characteristics, i.e., radius as a function of time, for three bubbles is shown in Fig. 1.3. As the bubble radius decreases, it is difficult to measure the radius due to the limited resolution of the camera. At zero radius the bubble collapses and a shock wave is expected to initiate and a high velocity jet of fluid emerges. The shock wave pressure at collapse can be estimated using the Rayleigh-Plesset equation [4] given by

$$R\ddot{R}+\frac{3}{2}\dot{R}^2=\frac{P_R-P_\infty}{\rho} \quad (1.1)$$

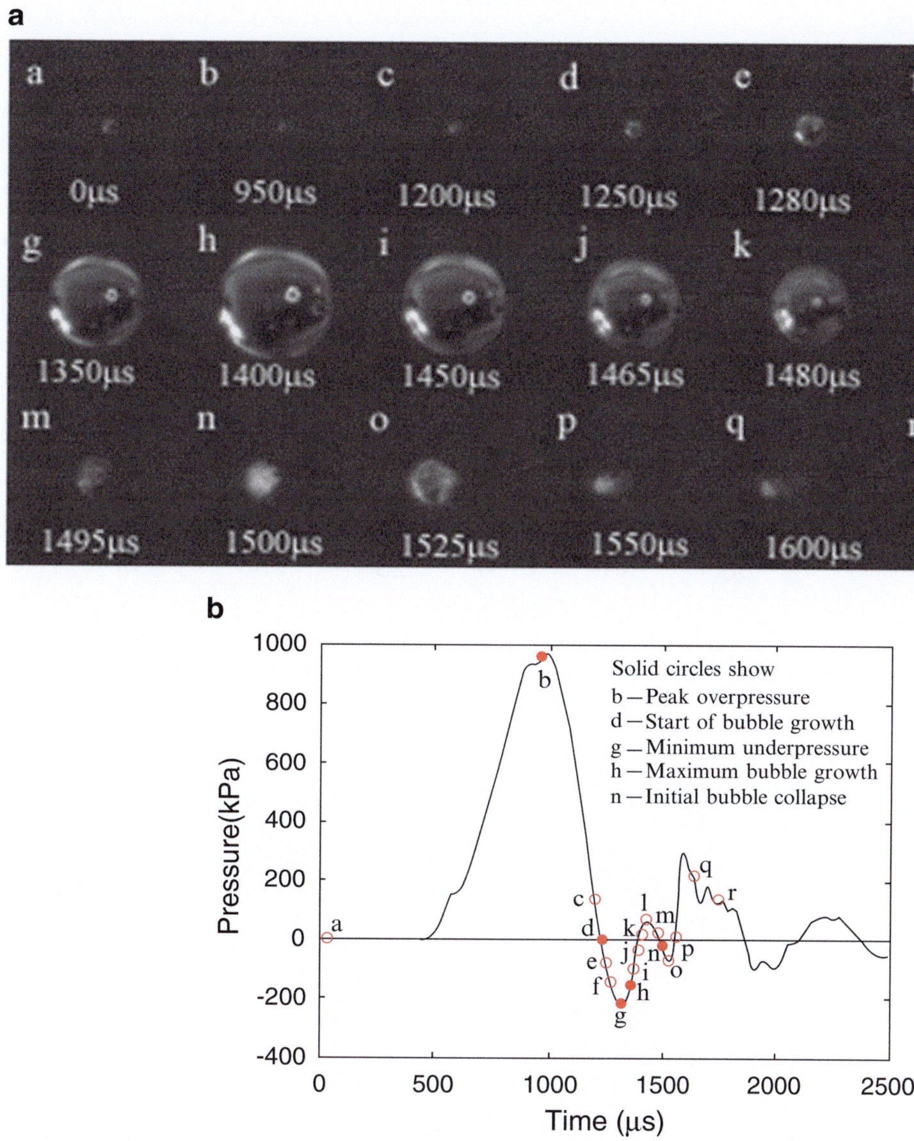

Fig. 1.2 Single bubble growth and collapse sequence. (**a**) Sequence of images showing evolution of a single bubble cavitation. (**b**) Corresponding pressure profile showing overpressure and underpressure phases

where R is the bubble radius, PR is the pressure in the liquid at the boundary of the bubble, P_∞ is the environmental pressure, and ρ is the density of surrounding fluid. For each bubble data shown in Fig 1.3, a fifth order polynomial was fit, i.e., $R(t)$. The time derivatives, $R.(t)$ and $\ddot{R}(t)$ were calculated. $P_\infty(t)$ was set as the far-field fluid pressure as measured by the pressure transducer (Fig. 1.2b). The pressure at the boundary of the bubble at each radius was calculated. Because the radius at the instance of bubble collapse (i.e., at $R=0$) is unknown, an exponential curve ($Pc=ke-R$) was fit for this data. By setting $R=0$, we determined the elusive collapse pressure Pc at the time of bubble collapse which is given by $Pc=k$. Typical collapse pressure for a large size bubble (i.e., large negative or tensile pressures) were as high as 100 MPa. For comparison, note from Fig. 1.2b that a typical value of incident blast overpressure was only around 1 MPa, and the blast under pressure was approximately 250 kPa. Thus the local pressure at the time of bubble collapse could be as high as 100 times the incident over pressure or approximately 400 times the blast under pressure. In our experiments, Pc was determined to be 29–125 times higher than the measured peak overpressure.

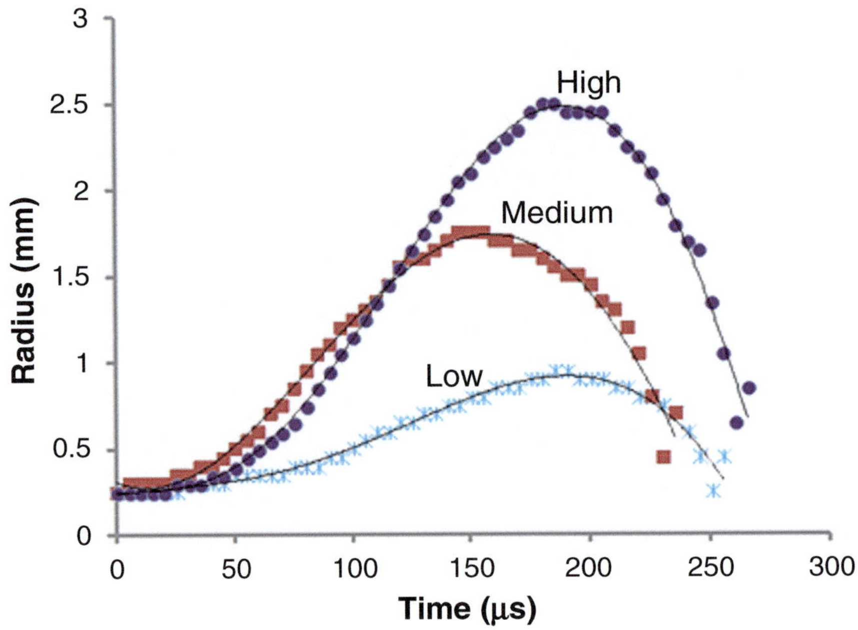

Fig. 1.3 Bubble time history for bubble growth and collapse during testing. Exponential equations were fit and pressure at $r = 0$ was estimated to be the bubble collapse pressure, *Pc*

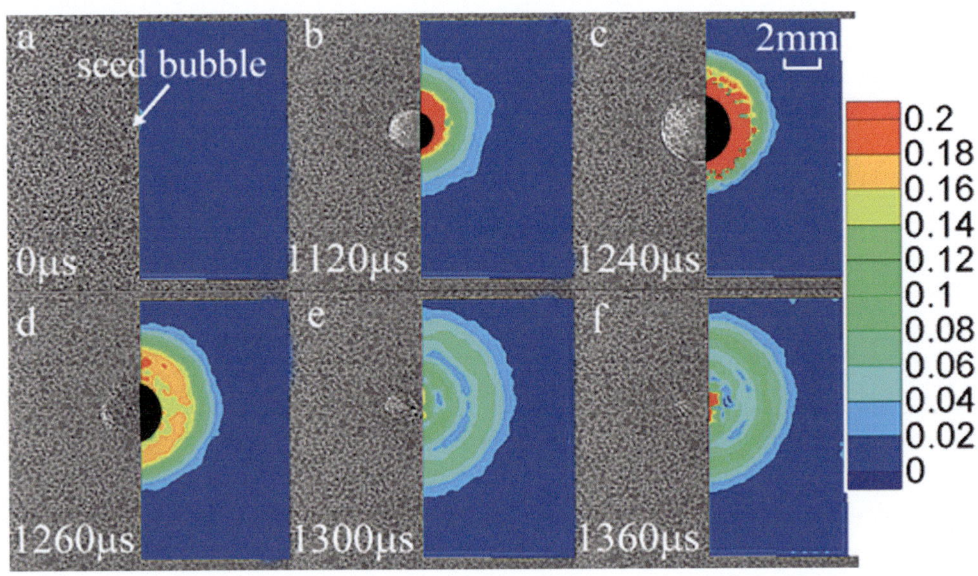

Fig. 1.4 Typical hydrogel deformation behavior for bubble cavitation underneath a speckled hydrogel slice. *Left half* of each of the high speed images show bubble growth (**b & c**) and collapse (**d & e**) at select time intervals and *right half* of each of the images show von Mises strain maps calculated using DIC

1.3.1 Hydrogel Deformation Due to Single Bubble Cavitation

To capture effects of the local shock wave generated with bubble collapse, a single seed bubble was placed underneath the hydrogel slice. Figure 1.4 shows a series of high speed images for a typical experiment during the bubble growth (b, c) and collapse (d, e) on the left side of each image, and the calculated strain field in the hydrogel at that moment on the right side of each image. Bubble growth was captured due to transparency of hydrogel. The von Mises strain field was determined to be radially symmetric, and a clear circumferential ring pattern evolved with time. The strain amplitude at the center of the bubble reached a peak value with maximum bubble radius (at c) and then decreased as the bubble collapsed (e). After the bubble collapsed (f), the local strain again increased at the center of the hydrogel. These large compressive strains are attributed to a localized shock wave originating from the collapsed bubble.

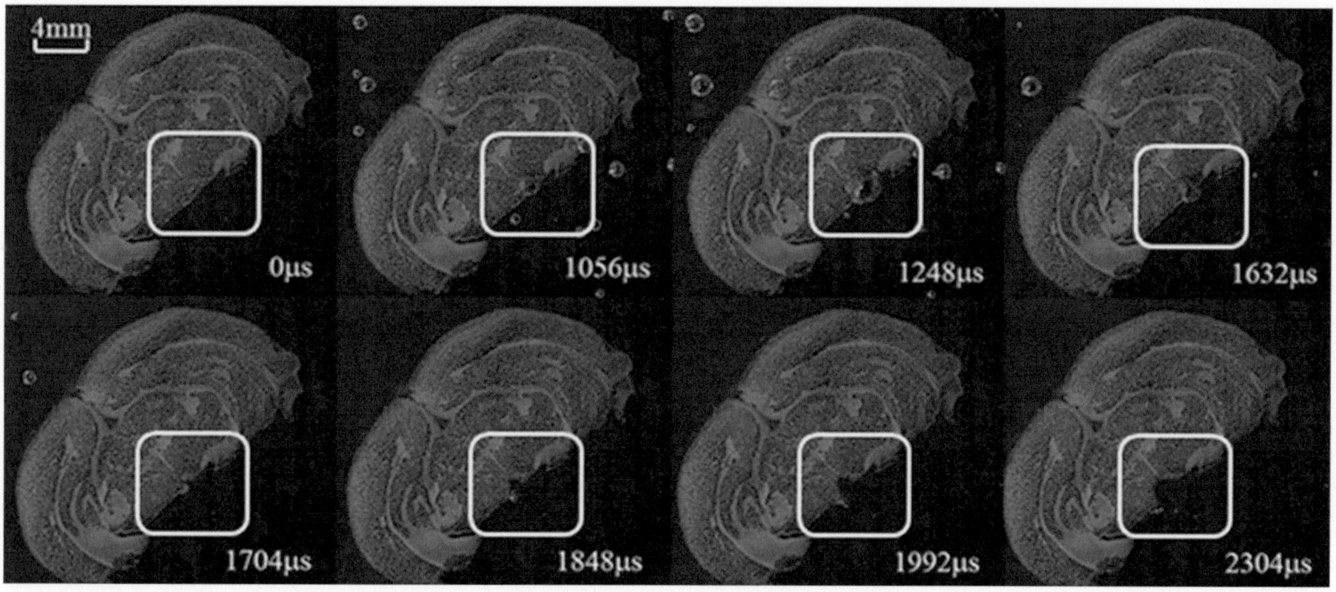

Fig. 1.5 Effect of cavitation on a rat brain tissue slice which was speckled. The images show bubble initiation (**b**), growth (**b**–**d**) and collapse (**f**). Cavitation-based damage with bubble collapse was observed in the hypothalamus region (*box*)

1.3.2 Cavitation Experiments with Brain Tissue

To appreciate the potential for cavitation to mediate blast injury, a brain tissue was introduced in the test cell. Figure 1.5 shows the effect of cavitation events on a rat brain tissue slice. Note that the maximum bubble size occurs at t = 1248 μs where the tissue tearing starts to occur well after the bubble has collapsed at time t = 1704 μs. These observations reveal the potential for collapse events to deform and injure soft brain tissues.

1.4 Conclusions

A novel system was developed for generating single bubble cavitation. The testing method was applied on a soft tissue surrogate. High speed imaging was used to capture bubble growth and elucidate cavitation effects. Cavitation effects resulted in far greater strains than seen with initial passage of the overpressure component. Future studies will further focus on testing effects of cavitation on brain tissue slices.

Acknowledgement This research was funded by the Army Research Office award number W911NF-10-1-0276.

References

1. Elder, G.A., Cristian, A.: Blast-related mild traumatic brain injury: mechanisms of injury and impact on clinical care. Mt Sinai J. Med. **76**, 111–118 (2009)
2. Rosenfeld, J.V., McFarlane, A.C., Bragge, P., Armonda, R.A., Grimes, J.B., Ling, G.S.: Blast-related traumatic brain injury. Lancet Neurol. **12**, 882–893 (2013)
3. Rayleigh, J.W.S.: On the pressure developed in a liquid during the collapse of a spherical cavity. Philos. Mag. **34**, 94–98 (1917)
4. Plesset, M.S., Chapman, R.B.: Collapse of an initially spherical vapour cavity in neighbourhood of a solid boundary. J. Fluid Mech. **47**, 283–290 (1971)
5. Ravichandran, G., Subhash, G.: Critical-appraisal of limiting strain rates for compression testing of ceramics in a split hopkinson pressure bar. J. Am. Ceram. Soc. **77**, 263–267 (1994)

Chapter 2
Inflation Test of the Human Optic Nerve Head Using Digital Volume Correlation

Dan E. Midgett, Harry A. Quigley, Mary E. Pease, Christian Franck, Jennet Toyjanova, and Thao D. Nguyen

Abstract Understanding the mechanical behavior of the Optic Nerve Head (ONH) is important for understanding the pathophysiology of glaucoma. We have developed an inflation test that uses second harmonic generation (SHG) imaging and digital volume correlation (DVC) to measure the deformation response of the lamina cribrosa, the connective tissue structure of the ONH, to controlled pressurization. Human eyes were obtained from a tissue bank. The episclera and extraocular tissues were removed, the optic nerve was cut flush with the sclera, and the eye, secured to a holder, was pressurized using a saline solution injected through the anterior chamber. Pressure was controlled using a manometer. A laser-scanning microscope was used to acquire SHG images of the deforming tissue structure, and the DVC algorithm was used to post-process the SHG volumes to calculate the 3D displacement field. We evaluated the baseline error and effect of creep by applying the DVC algorithm to sequential z-stacks acquired at the same pressure separated by different hold times. The method was applied to measure the deformation response to pressure elevations representative of glaucoma.

Keywords Digital volume correlation • Glaucoma • Lamina cribrosa • Biomechanics • Inflation

2.1 Introduction

In glaucoma, the progressive loss of the retinal ganglion cells (RGCs) in the ONH leads to permanent vision loss and eventual blindness. The intraocular pressure (IOP) is associated with the prevalence and severity of glaucoma damage, and the level of IOP is used to screen for the disease. Animal studies have shown that a chronic increase in IOP causes glaucoma damage, and lowering pressure slows the progression of damage in the majority of patients. However, there is a wide distribution of IOP among glaucoma patients and among those without glaucoma. While glaucoma patients on average have higher IOP, nearly half of those with glaucoma damage have lower than normal IOP. Moreover, a significant number of ocular hypertensives, with higher IOP, do not have glaucoma damage [1–4]. These observations suggest that the susceptibility of the ONH to glaucoma damage might depend on its mechanical properties of the ONH and surrounding tissues. Computational modeling studies of the posterior sclera have suggested that the elastic modulus of the LC and sclera strongly correlates with the strains experienced in the LC [5].

Both clinical and laboratory studies have measured differences in the mechanical behavior and structure of the posterior sclera, LC and ONH of glaucoma and normal eyes. Structural changes in glaucoma include the remodeling of the anisotropic collagen fiber and elastin structure at the lamina cribrosa [4] and axial elongation and thinning of the sclera [6–8]. Hommer et al. [9] measured an increase in ocular rigidity with the progress of OAG. Downs et al. [10] found a larger relaxation modulus for scleral strips from monkey eyes with experimentally induced early glaucoma. Girard et al. measured a stiffening of the sclera [11]. Coudrillier et al. [12] reported a thickening of the peripapillary sclera, slower circumferential creep rates, and a stiffer behavior in the meridional direction of the peripapillary sclera in human eyes with glaucoma. Few studies have measured the mechanical behavior of the lamina cribrosa. In this study, we present a method to measure the

D.E. Midgett (✉) • T.D. Nguyen
Department of Mechanical Engineering, Johns Hopkins University, Baltimore, MD, USA
e-mail: dmidget1@jhu.edu

H.A. Quigley • M.E. Pease
Wilmer Eye Institute, Johns Hopkins Medical Institute, Columbia, MD, USA

C. Franck • J. Toyjanova
School of Engineering, Brown University, Providence, RI, USA

S.A. Tekalur et al. (eds.), *Mechanics of Biological Systems and Materials, Volume 6*,
Conference Proceedings of the Society for Experimental Mechanics Series, DOI 10.1007/978-3-319-21455-9_2

three-dimensional deformation field in the lamina cribrosa under controlled inflation. The method uses laser-scanning second harmonic generation microscopy to image the collagen structures of the deforming volume of the lamina cribrosa and DVC correlation for deformation mapping.

2.2 Methods

2.2.1 Human Whole-Globe Eye Preparation

Five human donor eyes were received from the National Disease Research Interchange 24-h post-mortem. Eyes were stored at 4 °C and tested within 48 h. The sclera was cleaned of orbital tissues and the exposed nerve and episclera shaved off with a razor blade to expose the collagen beams of the lamina cribrosa. The remaining globes were then mounted in a fluid-filled reservoir to keep the tissues hydrated and held in place by three needles (Fig. 2.1a). A cannula was then inserted into the anterior chamber of the eye through the cornea to inflate the posterior segment, through injection of phosphate buffered saline (PBS) into the anterior segment. Pressure was controlled by the height of a water column connected to the cannula line. The baseline pressure was 20 mmHg.

2.2.2 Porcine Posterior Sclera Preparation

Four porcine eyes were received from Animal Technologies, stored at 4 °C, and tested within 48 h. The sclera was cleaned of orbital tissues and the exposed nerve and episclera shaved off with a razor blade to expose the collagen beams of the LC (Fig. 2.1b). The eyes were then glued using Permabond 910 to a polycarbonate ring, centered on the ONH and placed 3 mm anterior of the equator. The LC and posterior sclera were kept hydrated with water for 20 min as the tissue adhesive dried. The cornea and lens along with 2 mm of the surrounding anterior sclera was then removed using a scalpel and the aqueous humor, retina, and choroid were removed. The exposed scleral edge was then scored using a razor blade and the edges glued flat to the back of the holder to provide a watertight seal. The assembly was allowed to dry for an additional 20 min. The specimen was mounted on a custom inflation chamber and covered with PBS to ensure hydration (Fig. 2.1c). The specimen was aligned level to the imaging objective using the two tilt-correcting gears. The pressure was controlled by the height of a water column connected to the pressure line and initially set to a baseline pressure of 5 mmHg.

2.2.3 Imaging

The same imaging setup was used for inflation testing of both whole human globe and porcine posterior scleral cup. An inverted Zeiss LSM 710 NLO scope with a coherent Chameleon Ultra II laser was used to excite the exposed LC beams using backscattered Second Harmonic Generation (SHG) (Fig. 2.2). A 780 nm excitation was used to ensure a strong

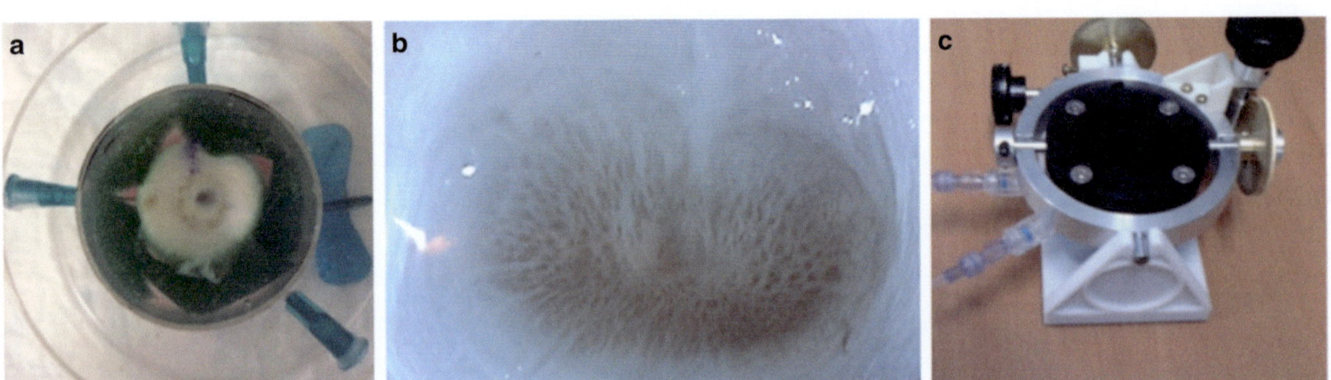

Fig. 2.1 Experimental preparation: (**a**) Exposed LC of prepared eye, (**b**) whole-globe human eye holder and (**c**) posterior sclera holder

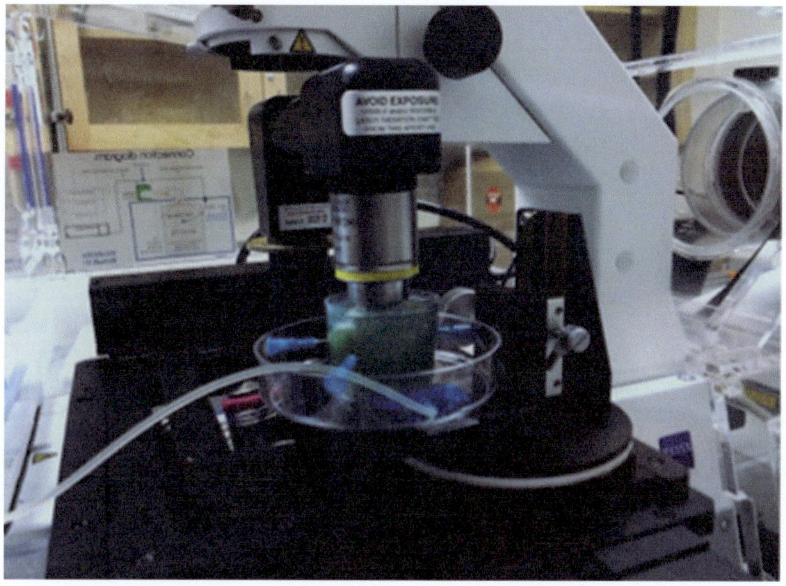

Fig. 2.2 Zeiss LSM 710 with whole-globe human eye

emission signal from the collagen [13] and the signal was collected using a short-pass 490 nm filter. The objective was centered on the central retinal artery and vein (CRAV). Images were acquired as a 2×2 tile using four 512×512 pixel images at 10× magnification. A 2×2 tiled z-stack was sized to cover most of the LC, which was approximately 2–3 mm in diameter, and the zoom factor was varied between 0.6 and 1 to adjust the field of view. The z-stacks were acquired with a short pixel dwell time (pdt) of 5 μs up every 5 μm starting at the distance closest to the choroid at which features were first visible (approximately 300 μm in depth) and up to the outer surface of the LC. A duplicate set of 2×2 tiled z-stacks was acquired immediately afterwards. The 2×2 tiled images were stitched using the Zeiss software to minimize discontinuities, and the data exported as a series of tiff files.

2.2.4 *Mechanical Testing*

The specimens were allowed to equilibrate to the baseline pressure for 45 min after which a series of images were acquired as described in the imaging protocol. For whole-globe specimens, additional sets of images were acquired after 60 and 75 min of equilibration to measure the creep. The pressure was increased to 45 mmHg, after which the specimen was allowed to equilibrate for an additional 45 min. The focus was readjusted to account for the average posterior displacement using a landmark within the tissue. This adjustment was recorded and added back to the DVC displacement calculations. The imaging protocol was repeated at the new pressure. For the porcine posterior scleral cup specimens, duplicate 2×2 tiled z-stacks were obtained 5, 10, and 45 mmHg. The specimen was allowed to equilibrate for 45 min between each pressure to minimize creep (Fig. 2.3).

2.2.5 *Digital Volume Correlation*

To calculate displacements, image sets were reconstructed into a three-dimensional matrix of image intensity in Matlab. The X and Y dimensions corresponded to the horizontal and vertical axes of each tiled image and the Z dimension corresponded to the anterior position within the LC. The Fast-Iterative DVC Algorithm developed by Toyjanova et al. [14] was used to correlate the image intensity of the reference and deformed volumes. A starting subset size of 128×128×64 pixels and a coarse calculation spacing of 32×32×16 pixels was selected for the first iteration. On the second iteration the subset size was halved to 64×64×32 pixels and the calculation spacing reduced to 16×16×8 pixels. Successive iterations of the FIDVC algorithm refined the spacing by an order of two at each iteration, but retained the same subset size until the field calculations were spaced at 4×4×2 pixels in XYZ for a total of four iterations. The resolution of our calculations was 1.6–2.5 μm/pixel in XY depending on the zoom factor used during imaging.

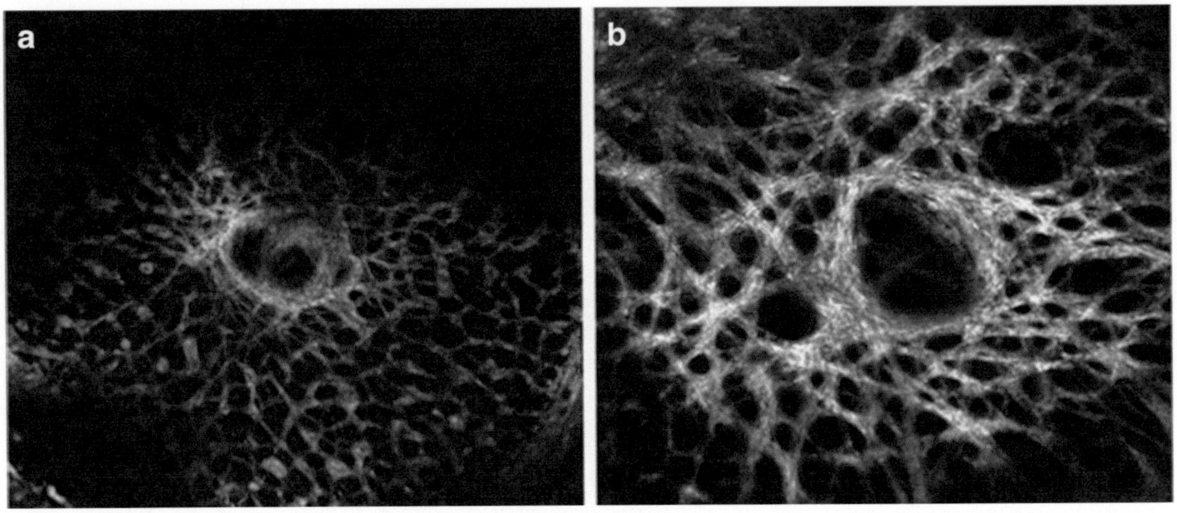

Fig. 2.3 Two 2×2 tiled second harmonic generation image of: (**a**) the human lamina cribrosa, and (**b**) pig lamina cribrosa

2.2.6 Error Measurements and Validations

To calculate the baseline displacement error and uncertainty, the duplicate 2×2 tiled z-stacks acquired back-to-back at each pressure were correlated. We defined the in-plane displacement magnitude as,

$$D = \sqrt{V^2 + U^2} \tag{2.1}$$

where U and V are the DVC displacements in the X and Y directions. The out-of-plane displacement in the Z direction is given by W. The displacements obtained for the duplicate volumes should be zero because the images were acquired at the same pressure, thus the DVC displacement field provides the baseline error under static conditions. We define the in-plane and out-of-plane displacement error as,

$$e_{xy} = \frac{1}{M}\sum_{x,y,z} D, e_z = \frac{1}{M}\sum_{x,y,z}\sqrt{W^2} \tag{2.2}$$

where M is the amount of calculation points in the volume.

In addition, we applied a virtual displacement field representing a rigid body displacement and an equi-triaxial stretch to one of the volumes using the matlab functions affine3d() and imwarp() and correlated the virtually deformed volume to its duplicate. The error between the DVC displacements and the applied virtual displacements was calculated at each DVC grid point. Points that had errors larger than 1 pixel were removed and the remaining region was used to calculate strains.

2.2.7 Strain Calculation

The validated displacement data was fit to a tri-linear function in a 7×7×7 cubic neighborhood about each calculation point. The displacement derivatives from the fit were then used to calculate the six Lagrangian strain components of the tensor E at each point. The Lagrangian strains were calculated as follows:

$$F = \begin{bmatrix} 1+U_{,x} & U_{,y} & U_{,z} \\ V_{,x} & 1+V_{,y} & V_{,z} \\ W_{,x} & W_{,y} & 1+W_{,z} \end{bmatrix}, E = \frac{1}{2}\left(F^T F - I\right) \tag{2.3}$$

2.3 Results

See Table 2.1

2.3.1 *Image Setting Optimization*

The baseline error and uncertainty calculations were used to select the optimum imaging settings for acquiring data with fresh human LC tissues. Acquisition speed was optimized by using three different acquisition speeds with specimen H1473 (Fig. 2.4), demonstrating that a fast pixel dwell time of 5 μs, produced lower baseline error than the slower dwell times, 40 μs and 80 μs. This suggested that significant creep occurred during the imaging time of the longer pixel dwell times. The baseline error also decreased with equilibrium time (Fig. 2.5). This is because creep rate decreased with time and less creep occurred during image acquisition after longer equilibration time. The mean absolute error for a 45-min equilibration time was subpixel, 0.4–0.8 μm, for all five human specimens acquired using the 5 μs pixel dwell time. Based on these results, specimens were equilibrated for 45 min, or longer when possible, to minimize the baseline error.

2.3.2 *Whole-Globe Human Summary*

Several issues were detected with the whole-globe human experiments. The specimen surfaces were often tilted resulting in large areas being out of focus resulting in low image intensity and poor contrast. A large baseline pressure was needed, at least 20 mmHg, for the globe to maintain shape. The human specimens also exhibited a large amount of rigid-body motion (RBM) indicating that the majority of the displacement was caused by rotation or translation of the eye during inflation. In-plane displacement magnitudes were 91.45 ± 35.28 μm and 68.61 ± 9.43 μm for the two normal eyes, and 48.34 ± 9.10 μm and 40.13 ± 24.93 μm for the two glaucoma eyes. The posterior displacement over the LC volume was 283.12 ± 5.53 μm and 296.46 ± 7.62 μm for the two normal eyes, and 177.84 ± 5.36 μm and 160.96 ± 6.29 μm for the two glaucoma eyes.

Table 2.1 Experimental parameters for human eyes

Eye ID	Glaucoma	Pixel dwell time (μs)	Magnification	Stitch pattern	Number of slices	Pressures (mmHg)	Equilibration times (min)	Acquisition time (min)
H1465	Yes	5	0.7×	2×2−512	61	20, 45	45, 60, 75	4
H1466	No	5	1.0×	1×1−512	41	20, 45	45, 60	2/3
H1470	No	5	0.7×	2×2−512	50	20, 45	45, 60, 75	3.5
H1471	Yes	5	0.7×	2×2−512	61	20, 45	60, 75	4
H1473*	No	5, 40, 80	0.7×	2×2−512	51	45	60, 67, 84	3.5, 8.5, 15

*This specimen was used to optimize the pixel dwell time and was not a pressure experiment

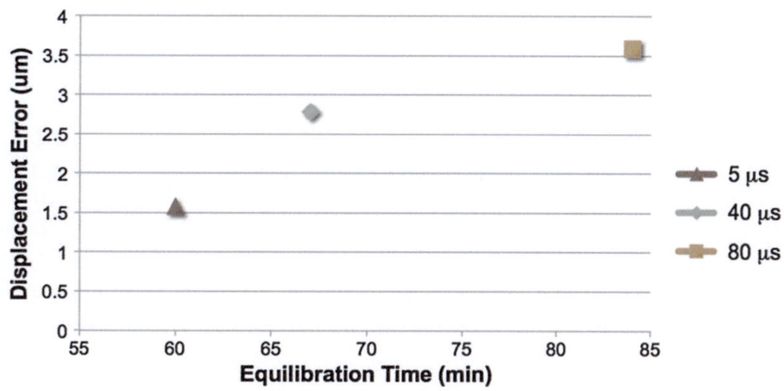

Fig. 2.4 Baseline error in XY-position (μm) at 45 mmHg for 3 different imaging pixel dwell times

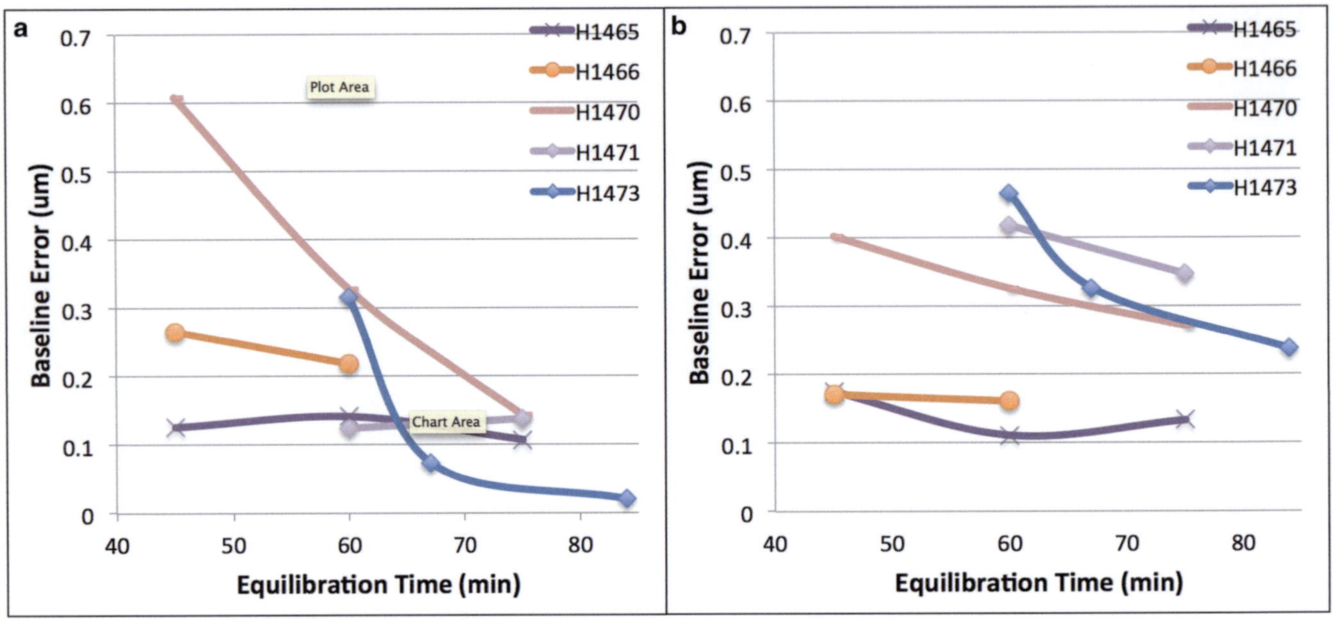

Fig. 2.5 Baseline error (μm) at 45 mmHg: (**a**) Baseline error in Z-position and (**b**) baseline error in XY-position

2.3.3 Porcine Posterior Scleral Cup

To address the issues experienced in the whole-globe human experiments, a custom inflation fixture was designed for inflation testing of excised scleral specimens. The fixture prevented rigid body motion of the specimen and allowed for manual tilt-correction to orient the surface of the specimen with the imaging plane. The following shows preliminary results for a specimen equilibrated at the lower baseline pressures of 5 mmHg and 10 mmHg.

2.3.4 Baseline Error

Figure 2.6 shows contours of the baseline error of the X, Y, and Z displacement components, averaged through all depths. There was a clear variation (about 0.4 μm) in the X-displacement between the right and left sides of the plot, indicating that creep probably occurred during imaging. The mean absolute error and standard deviation was 0.3518 ± 0.1184 μm for the in-plane XY displacements and 0.9693 ± 0.1305 μm for the out-of-plane Z displacements, which was lower than the resolution in those directions (2 μm/pixel in XY and 5 μm/slice in Z).

2.3.5 Virtual Validation

The correlation accuracy was evaluated using a virtually applied deformation field. Regions with high error were filtered out using Matlab. These regions included the central-retinal artery and vein, which is devoid of collagen, and another circular region shown in (Fig. 2.7). These areas were removed along with a 2000-μm ring at the scleral edge, where the denser collagen structure produced a saturated, low contrast image. The XY-averaged error as a function of the Z depth is shown in Fig. 2.8. The correlation error forms a U-shape as a function of depth, indicating that the volume correlates best in the middle Z-depths. Using this analysis, we excluded regions where the average error was above the imaging resolution (i.e. 2 μm in XY and 5 μm in Z), from strain calculation.

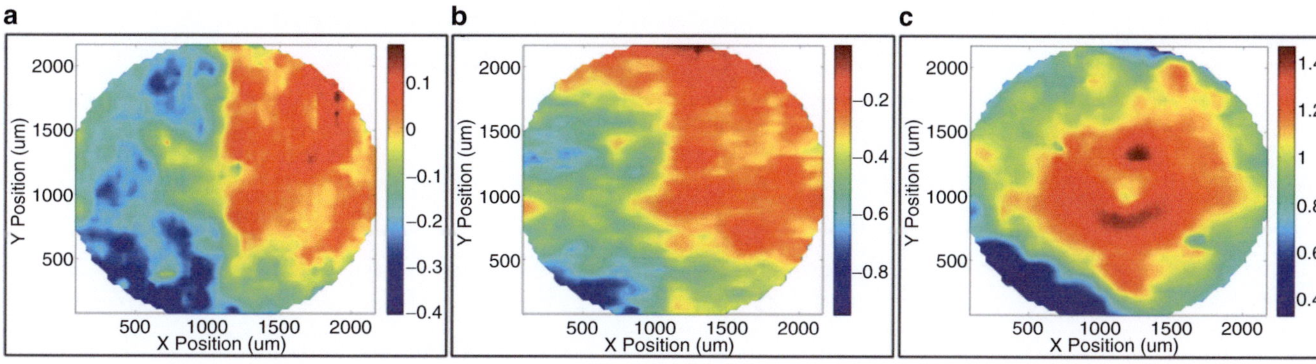

Fig. 2.6 Z-Averaged error plots (μm) at 45 mmHg: (**a**) X-displacement error, (**b**) Y-displacement error and (**c**) Z-displacement error

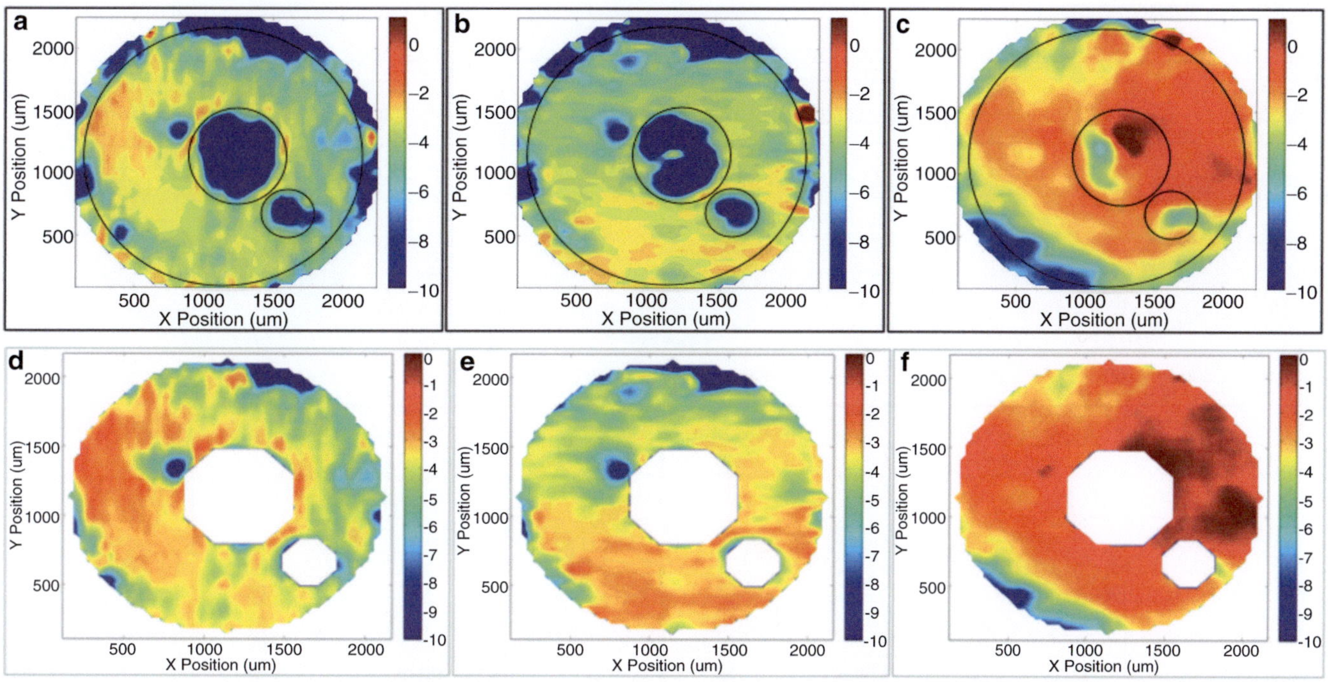

Fig. 2.7 Error (μm) to virtually applied displacement: (**a**) X-displacement error, (**b**) Y-displacement error, (**c**) Z-displacement error, (**d**) X-displacement error, filtered, (**e**) Y-displacement error, filtered and (**f**) Z-displacement error, filtered

2.3.6 Displacements and Strains

The X displacement field consisted of a stretch of 100 μm from left to right and symmetry about the CRAV. The Y-displacement was similar in form with a stretch of 70 μm from top to bottom indicating that this specimen was more compliant in the X-direction. The strain in the X-direction was 5 % on average, whereas the strain in the Y-direction was 3 % on average. The displacements and strains were consistent with the deformation of a membrane under inflation (Fig. 2.9).

2.4 Discussion

This study demonstrated that the natural contrast of collagen in the LC, imaged by SHG in an inflation experiment, could be correlated using DVC [14]. This allows for the calculation of 3D displacement and strain fields for fresh LC tissues and the comparison of tissue structure between different IOP levels. The displacement correlation error was calculated for static

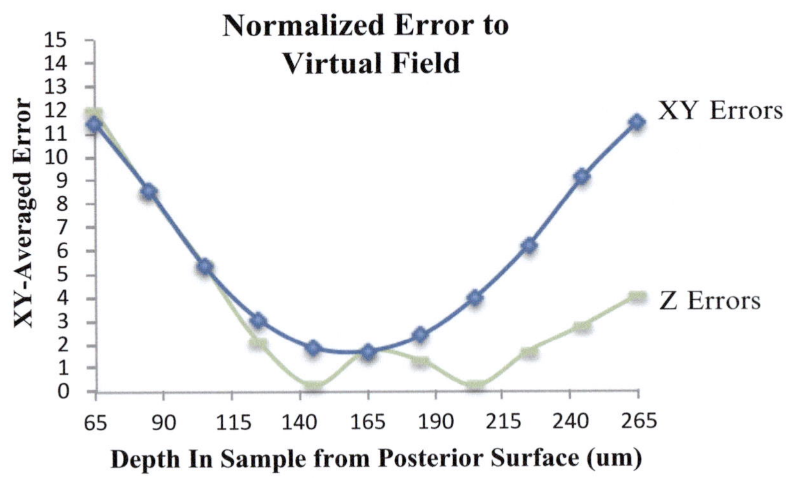

Fig. 2.8 XY-averaged error (μm) to virtually applied field

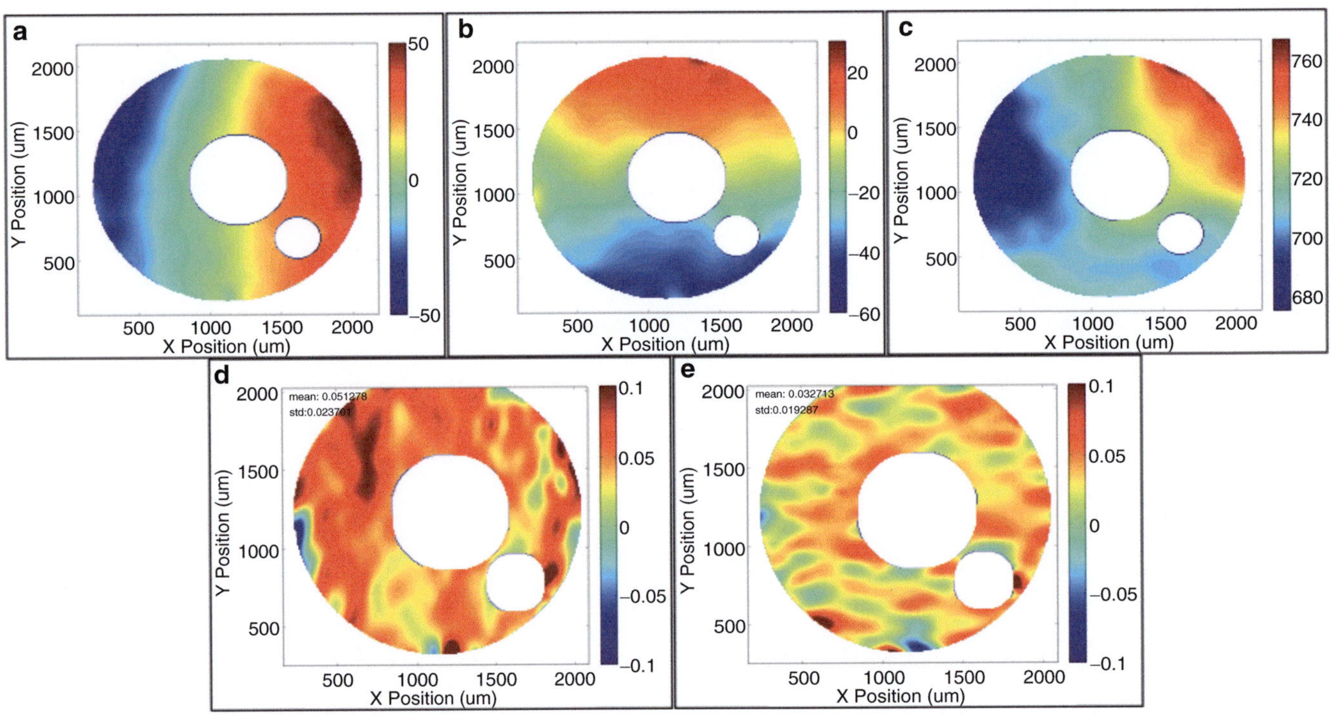

Fig. 2.9 Displacements (μm) and strains, 135 μm depth: (**a**) X-displacement, (**b**) Y-displacement, (**c**) Z-displacement, (**d**) E_{XX} and (**e**) E_{YY}

conditions and for an applied virtual deformation field that included uniform stretch and rigid body displacement. Baseline error during imaging was found to be sub-pixel for a pixel dwell time of 5 μs, an equilibration time of 45 min or more at each pressure, and an acquisition time of 4 min per volume.

References

1. Quigley, H.A.: Neuronal death in glaucoma. Prog. Retin. Eye Res. **18**, 39–57 (1997)
2. Burgoyne, C.F., Downs, J.C., Bellezza, A.J., Suh, J.K.F., Hart, R.T.: The optic nerve head as a biomechanical structure: a new paradigm for understanding the role of IOP-related stress and strain in the pathophysiology of glaucomatous optic nerve head damage. Prog. Retin. Eye Res. **24**, 39–73 (2005)
3. Morrison, J.C., Johnson, E.C., Cepurna, W., Jia, L.: Understanding mechanisms of pressure-induced optic nerve damage. Prog. Retin. Eye Res. **24**, 217–240 (2005)

4. Hernandez, M.R., Pena, J.D.O.: The optic nerve head in glaucomatous optic neuropathy. Arch. Ophthalmol. **115**, 389–395 (1997)
5. Sigal, I.A., Flanagan, J.G., Ethier, C.R.: Factors influencing optic nerve head biomechanics. Invest. Ophthalmol. Vis. Sci. **11**, 4189–4199 (2005)
6. Cone, F.E., Gelman, S.E., Son, J.L., Pease, M.E., Quigley, H.A.: Differential susceptibility to experimental glaucoma among 3 mouse strains using bead and viscoelastic injection. Exp. Eye Res. **91**, 415–424 (2010)
7. Nguyen, C., Cone, F.E., Nguyen, T.D., Coudrillier, B., Pease, M.E., Steinhart, M.R., Oglesby, E.N., Jefferys, J.L., Quigley, H.A.: Studies of scleral biomechanical behavior related to susceptibility for retinal ganglion cell loss in experimental mouse glaucoma. Invest. Ophthalmol. Vis. Sci. **54**, 1767–1780 (2013)
8. Cone-Kimball, E., Nguyen, C., Oglesby, E.N., Pease, M.E., Steinhart, M.R., Quigley, H.A.: Scleral structural alterations associated with chronic experimental intraocular pressure elevation in mice. Mol. Vis. **19**, 2023–2039 (2013)
9. Hommer, A., Fuchsjäger-Maryl, G., Resch, H., Vass, C., Garhofer, G., Schmetterer, L.: Estimation of the ocular rigidity based on measurement of pulse amplitude using pneumotonometry and fundus pulse using laser interferometry in glaucoma. Invest. Ophthalmol. Vis. Sci. **49**, 4046–4050 (2008)
10. Downs, J.C., Suh, J.K.F., Thomas, K.A., Bellezza, A.J., Hart, R.T., Burgoyne, C.F.: Viscoelastic material properties of the peripapillary sclera in normal and early-glaucoma monkey eyes. Invest. Ophthalmol. Vis. Sci. **46**, 540–546 (2005)
11. Girard, M.J.A., Suh, J.K.F., Bottlang, M., Burgoyne, C.F., Downs, J.C.: Biomechanical changes in the sclera of monkey eyes exposed to chronic IOP elevations. Invest. Ophthalmol. Vis. Sci. **52**, 5656–5669 (2011)
12. Coudrillier, B., Tian, J., Alexander, S., Myers, K.M., Quigley, H.A., Nguyen, T.D.: Biomechanics of the human posterior sclera: age- and glaucoma-related changes measured using inflation testing. Invest. Ophthalmol. Vis. Sci. **53**, 1714–1728 (2012)
13. Cox, G., Kable, E., Jones, A., Fraser, I., Manconi, F., Gorrell, M.D.: 3-Dimensional imaging of collagen using second harmonic generation. J. Struct. Biol. **141**, 53–62 (2003)
14. Bar-Kochba, E., Toyjanova, J., Andrews, E., Kim, K., Franck, C.: A fast iterative digital volume correlation algorithm for large deformations. Exp. Mech. **55**, 261–274 (2015)

Chapter 3
In Situ NANO-Indentation of Round Window Membrane

Hirobumi Watanabe, Anil K. Lalwani, and Jeffrey W. Kysar

Abstract The round window membrane (RWM) is the terminal boundary between the fluid-filled inner ear and air-filled middle ear where the sound wave is released from the inner ear after exciting the neuronal sound transducers in the basilar membrane. For the treatment of inner ear diseases, the RWM is an attractive entrance to introduce therapeutic reagents by producing micro-scale perforations. Therefore, the mechanical properties of this collagen-fiber-reinforced membrane are critical in understanding the functional role in hearing and engineering a device for drug delivery.

In this study, NANO-indentation was performed to determining the rupture criteria of freshly harvested RWMs of guinea pigs *in situ*. Tungsten probe tips with 20 and 5 curvature radius and 2 mm length were used to obtain load-displacement curves until rupture. As the tip size decrease, the rupture force decreased approximately from 5 to 0.3 mN. The Young's modulus of the membrane during the deformation and the stresses within the membrane at the rupture were analyzed inversely via finite element modeling. Biological variation of the mechanical properties was shown. These findings facilitate modeling of sound propagation within the inner ear in a normal condition and with a prosthetic device and developing micro-needles for drug delivery.

Keywords Nano-indentation • Biomedical engineering • Round window membrane • Rupture criteria • Hearing

3.1 Introduction

Ear is one of the most sophisticated sensory organ that constitutes mechanical systems evolved to capture the sound coming through the air and transduce the mechanical vibration into neural signals. The dynamic sound pressure travels from the outer ear into the inner ear through the oval window and will finally be released through a delicate membranous window, called round window membrane (RWM). The RWM must be substantially large and compliant enough to release the sound pressure effectively. At the same time, it must be strong enough to bear the quasi-static pressure fluctuation in the inner ear, to keep the perilymph from leakage and to defend the inner ear from infections. The RWM has been the route to gain access to the inner ear for clinical interventions such as cochlear implantation for electrical stimulation of neurons and direct mechanical amplification with a middle ear implant. Furthermore, the RWM is a candidate as a route to deliver therapeutic reagents [18]. Currently, our group aims at improving the efficacy of the drug delivery via introduction of micro perforations using micro needle array (Fig. 3.1). Here, designing a micro-needle array requires clear understanding of the mechanical properties to ensure that the successful perforation without mechanical failure of the micro-needles. We aim at designing a micro-needle that is capable of perforating the window with precise control over the size to restrict the exposure to the viral or bacterial infection into the inner ear. However, the mechanical properties of the RWM is poorly understood in order to model any of the natural phenomena nor clinical interventions. Especially, the penetration of the membrane with a microneedle involves the large deformation and rupture of a collagen fibers. Understanding this physical system calls for a proper experiment and numerical modeling.

H. Watanabe, Ph.D. (✉) • J.W. Kysar, Ph.D.
Columbia University School of Engineering, 220 Mudd,
500 West 120th Street, New York, NY 10020, USA
e-mail: hw2420@columbia.edu

A.K. Lalwani, M.D.
Columbia University Medical Center, 180 Fort Washington Avenue, Harkness Pavilion Bldg.,
8th Floor, New York, NY 10032, USA

S.A. Tekalur et al. (eds.), *Mechanics of Biological Systems and Materials, Volume 6*,
Conference Proceedings of the Society for Experimental Mechanics Series, DOI 10.1007/978-3-319-21455-9_3

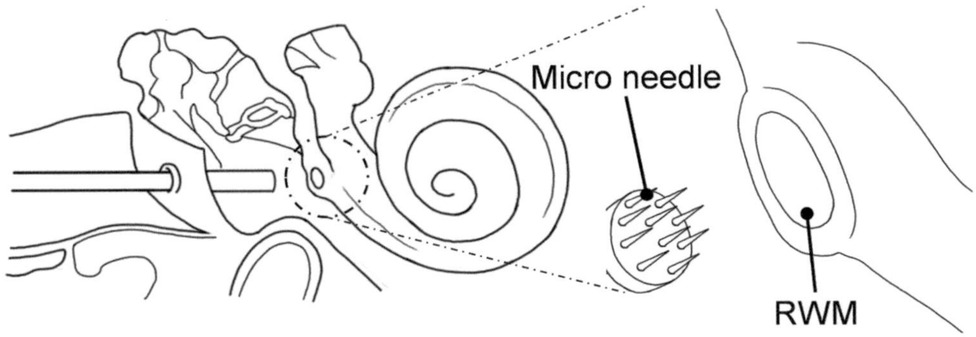

Fig. 3.1 Schematic diagram of RWM micro perforation

The RWM is a three-layered structure consisting of an outer epithelial layer that faces the middle ear, a central connective tissue layer, and an inner epithelial layer interfacing with the scala tympani. The connective tissues in the central layer contains fibroblasts, collagen, and elastic fibers. This layer provides mechanical strength to the RWM to bear the perilymph pressure [1]. As to the strength of the RWM, a previous study of the mechanical properties of human RWM until rupture performed in 1995 did not employ the proper physical model to analyze the experimental data [2]. Some of studies postulates that barotrauma at the pressure of 200 MPa or whiplash injury are postulated as causes of RWM rupture [3, 4]. As a primitive physical model of the whiplash injury, the pressure change across the RWM was modelled as the inertia of the perilymph produced from the rapid head movement [5]. However, due to the location of the RWM, the inspection of the RWM is uncommon compared to detection of the tympanic membrane rupture. Except for the exploratory tympanostomy, it is not possible to observe the actual state of the RWM. Therefore, more evidence would be desirable to conclude the strength of the RWM. When it comes to dynamic sound load on the RWM, though a more recent study [6, 7] showed very nice estimate of the linear elastic modulus, the use of the results are limited to small deformation (nm range) and high frequency vibration. Further, improving the understanding on the mechanical properties of the RWM will benefit clinical cases involving active middle ear implant [8], pathological conditions such as otitis media and Meniere's disease, [9, 10] and the intracochlear pressure changes that may have possibilities of RWM rupture or hearing decline [11].

Determination of the mechanical properties of soft tissues within the ear has been performed predominantly for the understanding of the biophysics in hearing. Indentation method was used to determine the mechanical properties of the tympanic membrane under the quasi-static load change (below 10 Hz) [12]. Recent researches on the constitutive modeling of tympanic membrane have shown progress by the improvement of the experimental technologies and the computational power [13–17]. Within the inner ear where the dynamic sound is transduced to neural signal, quite a few number of researchers have included the complex structure of the underlining microscopic tissues within the basilar membrane and the hair cell motion [18]. To further advance these understanding toward the goals of improvement of clinical practice and medical technologies, further investigation is necessary via different types of experiments and improvements in the modeling.

Nanoindentation with micro to nano probe tips is one of the ideal method to study the penetration of the RWM for the design of a microneedle. Especially, the extreme sensitivity of the nanoindenter is suitable to study the interaction of the RWM with an indenter probe tip with curvature radius less than 1 μm. Because the diameters of the collagen fibers that bear the load are a few hundreds, the extra sensitivity of the nanoindenter is beneficial to determine the small force [19]. Precise determination and elegant simplification of the geometry of the RWM are fundamental steps in the numerical analysis for the determination of the mechanical properties [20]. A micro computed tomography (μCT) with the resolution down to 1 μm is non-destructive method that will improve the current resolution of the anatomy more than 10-fold compared to the previous publications via micro MRI or OPFOS [21]. This detailed anatomy is vital for the modeling such as thickness variations and the effect of the boundary between the soft tissues with bone.

In this study, we studied the mechanical properties of the RWM. First, an experimental system was developed to determine the load-displacement curve when a RWM of guinea pigs *in vitro* was under highly localized load imposed by a microneedle with varied tip size until rupture using nano-indenter. Second, the perforated RWMs by the microneedles were examined using electron microscopy to determine the size of the perforations. Third, as a geometrical data for the finite element modeling, in addition to the data-set from our previous μCT scan and white light laser interferometry study, an improved μCT scan was performed to examine the thickness variation of a RWM. Lastly, all of these data-sets were used for the inverse analysis via finite element modeling to determine Young's modulus as well as the rupture criteria.

3.2 Method

Sample from Experimental Animals Adult Hartley guinea pigs of mixed gender weighing between 280 and 380 g were utilized in this study. The study protocol was approved by the IACUC of the Columbia University and met the guidelines of the National Institutes of Health. Each animal was sacrificed under euthanasia and the two cochlae from both ears were trimmed with a surgical drill within 1 h. The cochlae were trimmed down to expose the outer and inner surface of the RWM to air with enough moisture during the indentation (Fig. 3.2). The attachment of the RWM to the bone was kept intact to maintain the boundary conditions and the assumed pre-strain within the RWM. During trimming, the RWMs were always immersed in a saline solution. The cochlae bone were fixed to holes in a block of epoxy fiber glass laminate breadboard using UV curable dental composite (Fig. 3.3). The block and fixation with dental composite provided a rigid substrate during the indentation test and an inert material during the chemical process for the SEM observation. The RWM samples embedded in the aluminum block were stored in cold saline solution until indentation experiment.

NANO-Indentation Indentation tests were performed with G200 Nanoindenter (Agilent, TN). Figure 3.4 shows the schema of the experimental stage with a humidity control. Within the nano-indentation chamber, the moisture of the RWM was maintained using the custom made stage equipped with an inlet of humidity connected to a humidifier and an outlet of air driven by a vacuum cleaner. High humidity only within the sample holder pit was confirmed before the experiment using a humidity sensor CHS-MSS (TDK, Japan, Tokyo). First, a metal probe (Fig. 3.5) was lowered down onto a RWM surface until it makes a contact. The metal probe movement is controlled by continues stiffness mode (CSM) within the NANO-indentation system such that it moves with a constant speed (1 μm/s) with sinusoidal oscillation. The contact of the probe and the surface is determined by the phase change of the sinusoidal oscillation. The surface contact procedure was optimized

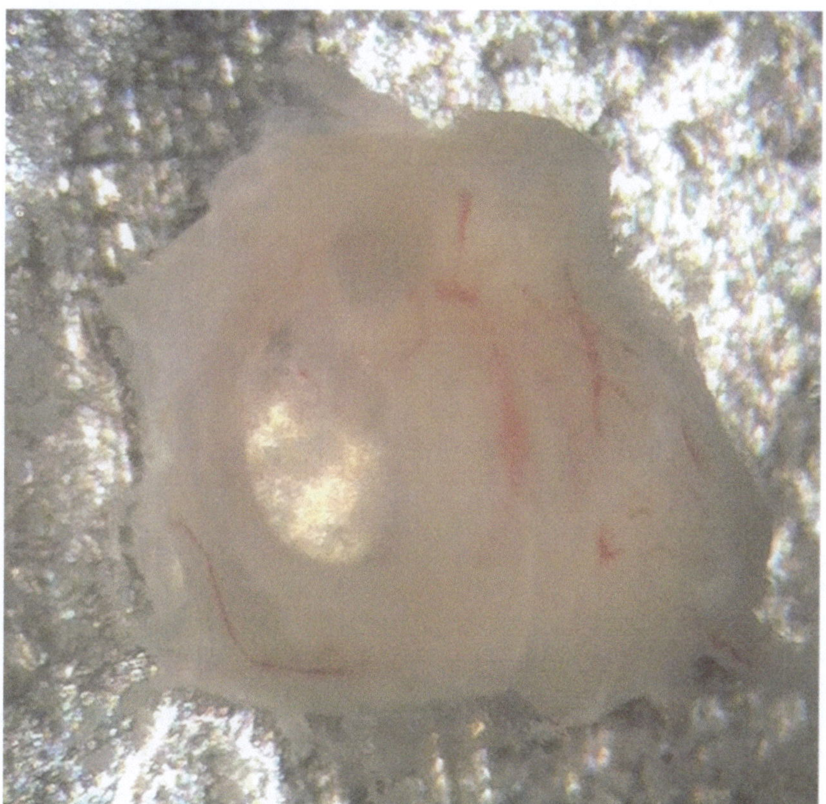

Fig. 3.2 A trimmed cochlea with a translucent RWM. The attachment of the RWM to the bone is intact

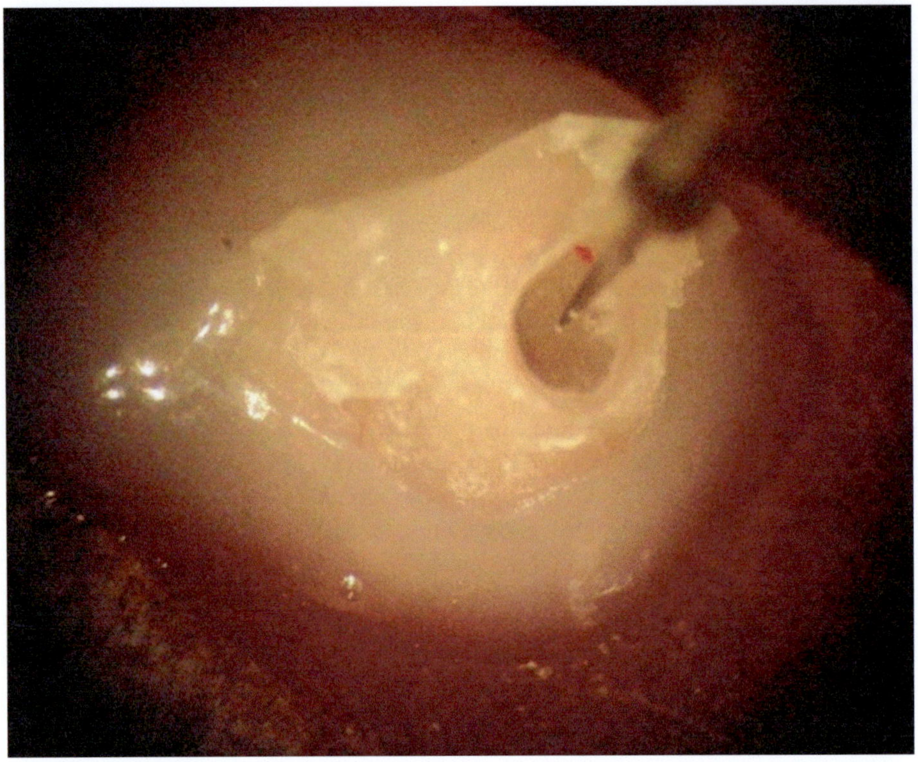

Fig. 3.3 The RWM embedded in a hole in a substrate block with a dental composite. Since the sample can't be visualized during the indentation within the nano-indentation chamber, this picture was taken with a micro needle in contact with the RWM for explanation

depending on the probe tip curvature. The metal probe was further brought down into the RWM until penetration. The penetration was confirmed automatically by the sudden decrease of the load. The curvature radii of 5 and 20 μm were used (Fig. 3.5). 5 μm Minutien Pins (Fine Science Tools, CA) was purchased. 20 μm probe tip was fabricated via electrochemical etching. These probe tips were mounted on the NANO-indenter tip holder using Crystalbond 509 (Structure Probe, PA).

SEM After indentation, scanning electron microscopy (SEM) was performed after dehydration and super-critical point drying. The detailed procedure was written in a previous publication [22]. Briefly, the sample was undergone dehydration process using a series of graduated concentrations of ethanol/water. After replacing water molecule with ethanol, hexamethyldisilazane (HDMS) reagent grade, ≥99 % (Sigma-Aldrich) was used to protect the tissue sample from drastic change of the surface tension as an effective replacement of critical-point drying. The dried sample was coated with gold and examined using a SEM, S-4700 (Hitachi, Japan).

Micro CT Previously using micro CT and Zygo, we have determined the size and shape of the RWM [22]. The data sets in this previous study were used. The shape of the RWM was an oval of major and minor axes: 1.29, 0.95 mm from plan view and had a saddle shape with curvature radii of 0.70 and −0.44 mm. We have scanned a RWM with a μCT scanner (1172 Skyscan, Belgium) at the exceptionally improved resolution of 1 μm. This scan data was used to confirm the thickness variation within the RWM. Independently from the indentation experiment, a trimmed guinea pig sample was fixed using neutral buffered 10 % formaldehyde (Sigma-Aldrich) for overnight. The sample was washed with tap water and μCT scan was performed. The typical duration was 5 h. To prevent dehydration, the sample was stored in a sealed tube during the scan.

Finite Element Modeling Inverse analysis was performed using finite element modeling software: ABAQUS in conjunction of Python code to access ABAQUS output data under the control of MATLAB. Figure 3.6a shows the simulation of the axis-symmetric model of nanoindentation of RWM with the radius of 0.5 mm under a contact of a 20 μm probe tip used in this study. At the rim of the disk, neither displacement nor bending were allowed. Mesh density was optimized by distributing more densely around the contact with the probe tip to allow smooth stress distribution as well as the rim of the disk to allow smooth bending. The probe tip was displaced downward up to 200 μm. Axis-symmetric model was used for the analy-

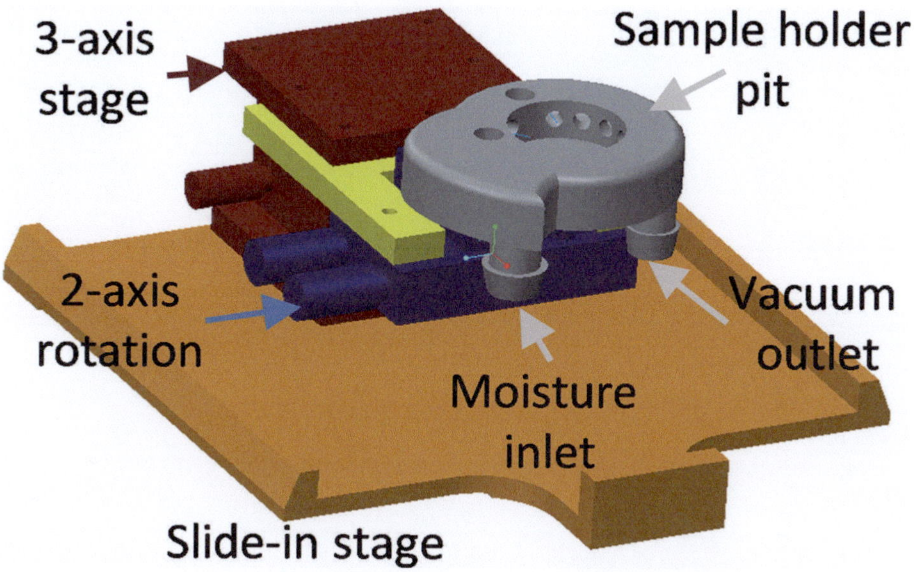

Fig. 3.4 The design of the custom made nano-indenter stage with humidity control and 5-axis control

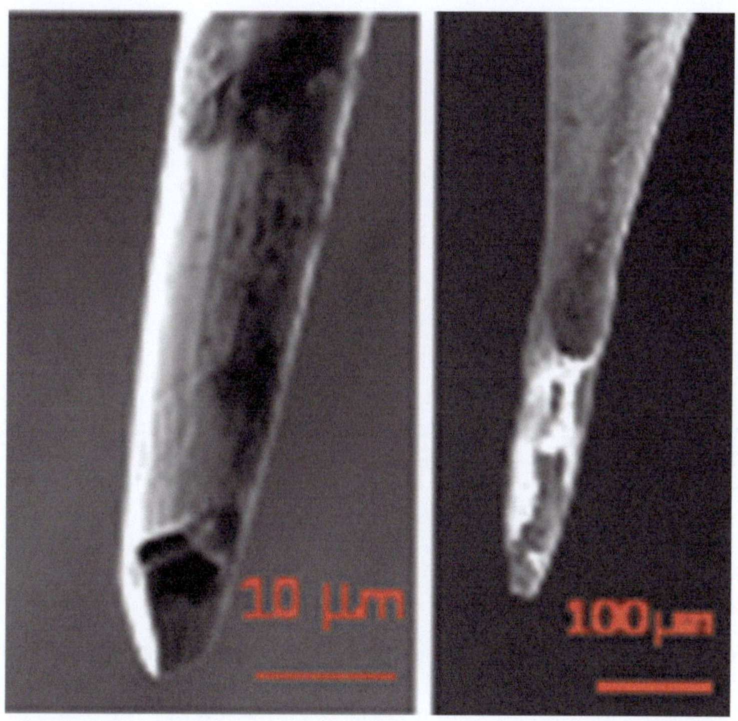

Fig. 3.5 Tips with the curvature radius of 5.0 and 20.0 μm

sis of the stress distribution in out-of-plane direction of the membrane in addition to radial direction. Even though we identified the shape of the RWM as more complex structure and simplified as a saddle shape with an oval in the plan view, the axis-symmetric model was used for this initial study for three reasons. First, the stress status of the RWM in the vicinity of the needle contact predominantly depends on the highly localized region because, roughly, the tensile stress decrease inversely proportional to the distance from the point of contact. Second, the contribution of the saddle shape predominantly comes from the concavity with the curvature radius of −0.44 mm (Fig. 3.6b). Thus, modeling the RWM as an axisymmetric concave will provide a good approximation of the contribution of the saddle shape to the indentation experiment. Under a

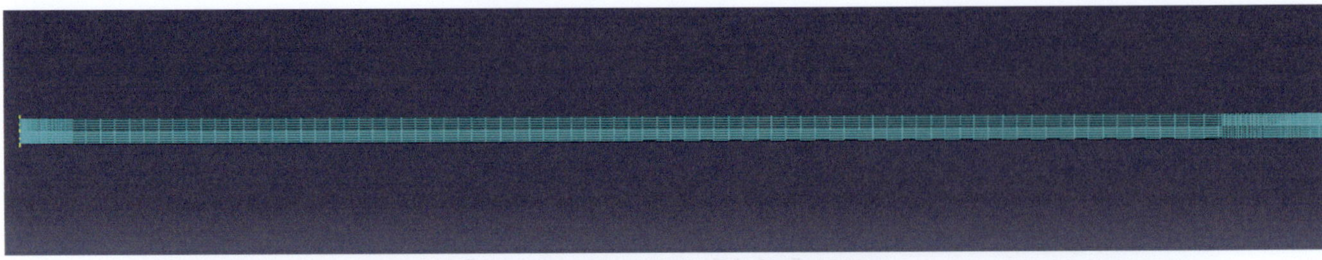

(a). An axis-symmetric model of a flat RWM

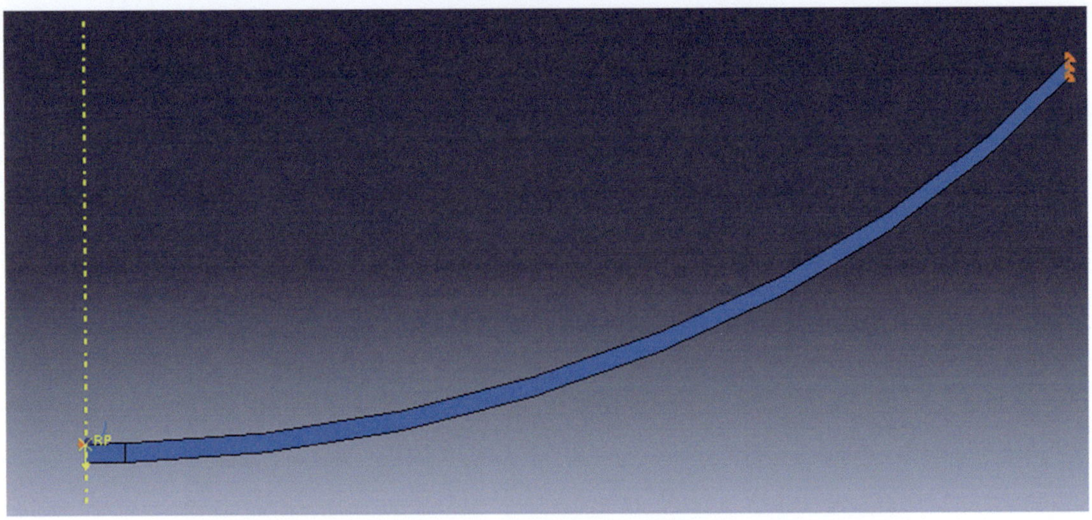

(b). An axis-symmetric model of the RWM with concavity

Fig. 3.6 ABAQUS simulation of the indentation of RWMs with a 20 μm probe tip

linear, isotropic assumption, Young's modulus was determined via the inverse analysis as the value that minimize the difference between the experimental data and the ABAQUS simulation for a flat disk with the thickness of 14 μm as well as a parabolic concave disk with the curvature radius of −0.44 mm at the center.

3.3 Results

NANO-Indentation Figure 3.7 shows a typical load-displacement curve obtained with 20 and 5 μm curvature tips. The curve obtained with 20 μm probe shows more than 5 mN load and 200 μm displacement at the time of penetration. The diameter of the rupture was about 20–50 μm. The 5 μm tip penetrated the RWM at <0.3 mN load and 60 μm displacement. Table shows the variance of the displacement and load at the penetration via 20 μm and fine probe tips. Both fine and 20 μm tips showed greater variance in displacement than load.

SEM Figure 3.8 shows a SEM image of one of RWMs with one holes created using a probe with curvature diameter of 20 μm. The positions of the hole were confirmed as deviation within 5 % from the center. Figure 3.9 shows a hole was created by an indentation procedure in which the probe was advanced 50 μm after the confirmation of the sudden drop of the load i.e., the rupture of the RWM. Figure 3.10 shows three holes created serially with an indentation procedure that prompted immediate retraction of probe tip after the rupture. In Figs. 3.11 and 3.12, the cochlea and the RWM penetrated with a 10 μm needle are shown.

Micro CT Figure 3.13 shows one section of the micro CT scan at the resolution of 1 μm. Red indicates the soft tissue structure and white indicates the hard structures. The thickness variation of the RWM was confirmed as 14–16 μm and the

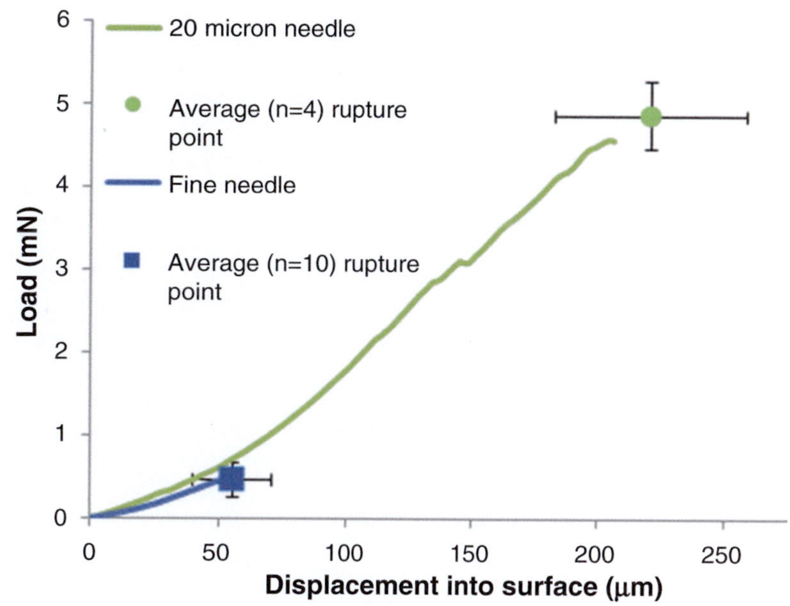

Fig. 3.7 NANO-indentation with tips with 20 and 5 μm in diameter

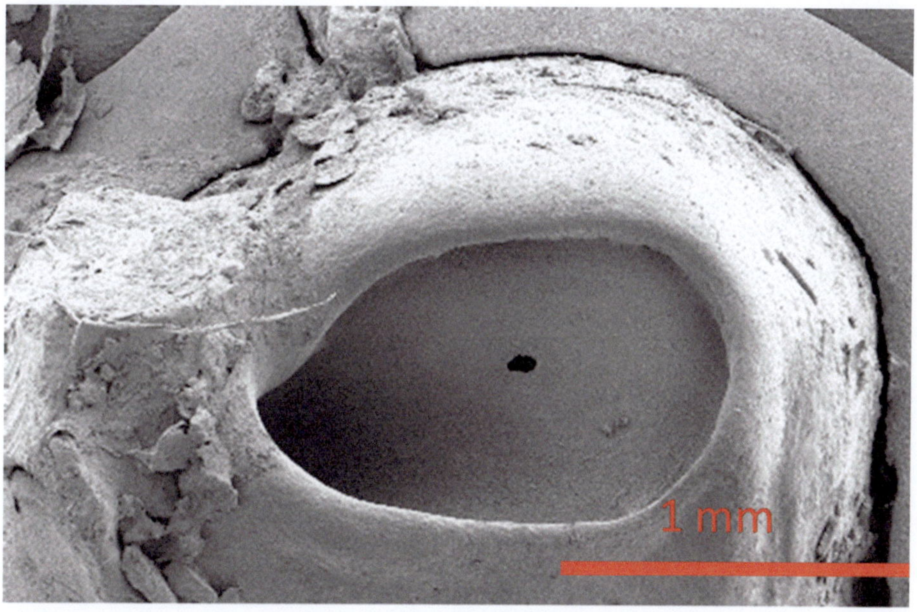

Fig. 3.8 Scanning electron microscope image of the RWM after punctuation with 20 μm tip

thickness variation was also quite small across the whole membrane except for a small portion of RWM. The attachment of the RWM to the bony round window sulcus was confirmed in this analysis. The thickening of the RWM was seen on the bony sulcus or outside of the region of the free standing membrane.

Finite Element Modeling Figure 3.14 shows the results of inverse analysis showing the experimental load-displacement curve up to 80 μm obtained using 20 μm probe tip, calculated curve with a flat disk and a disk with concavity of 0.44 mm in curvature radius. The best fit curves were obtained at the Young's modulus of 32.8 and 19.4 MPa with the flat disk and curved disk respectively. Clearly, the load-displacement curve with the consideration of the concavity shows the good fit with the

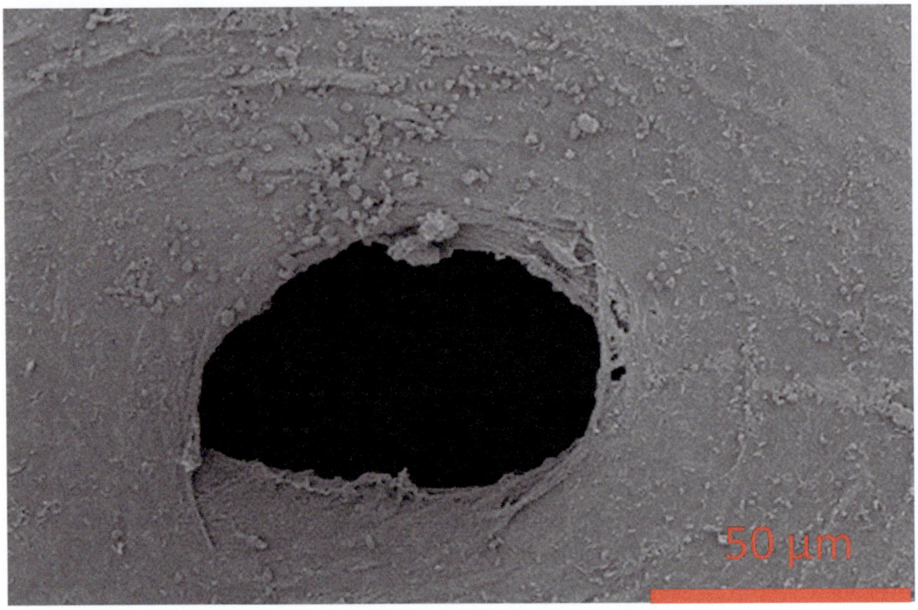

Fig. 3.9 Scanning electron microscope image of the RWM after punctuation with 20 μm tip

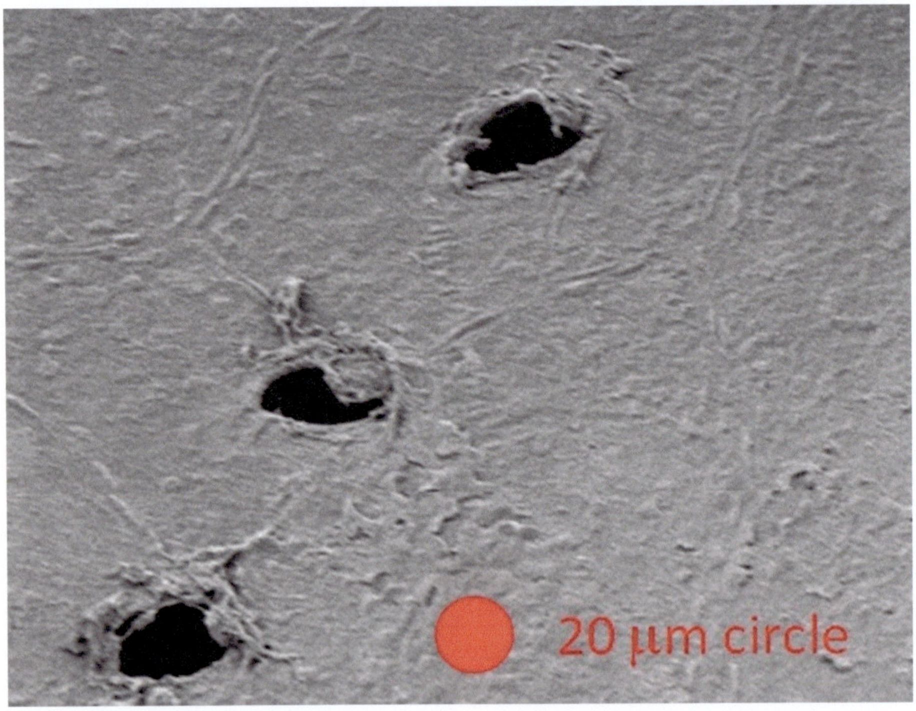

Fig. 3.10 Scanning electron microscope image of the RWM after punctuation with 20 μm tip

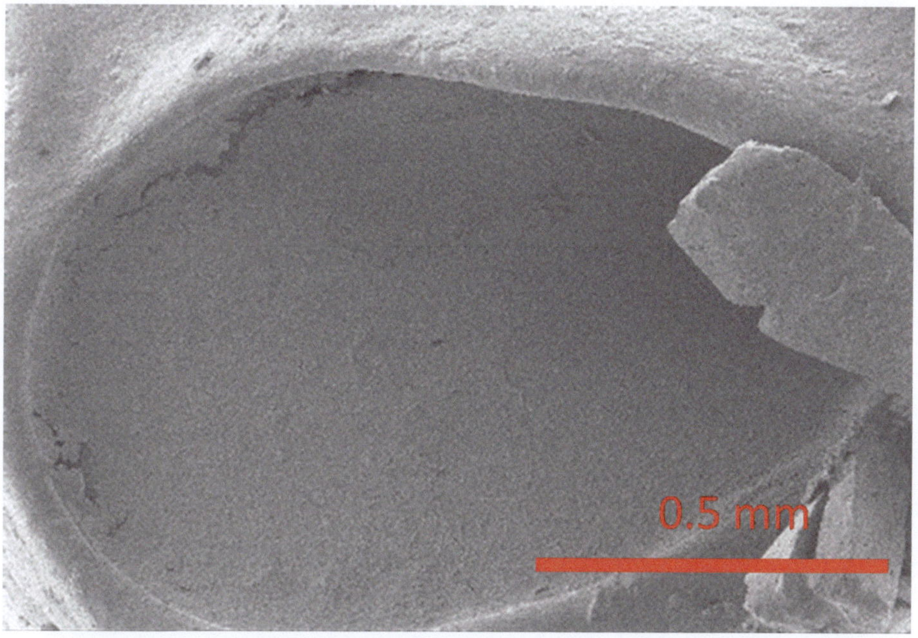

Fig. 3.11 Scanning electron microscope image of the RWM after punctuation with 20 μm tip

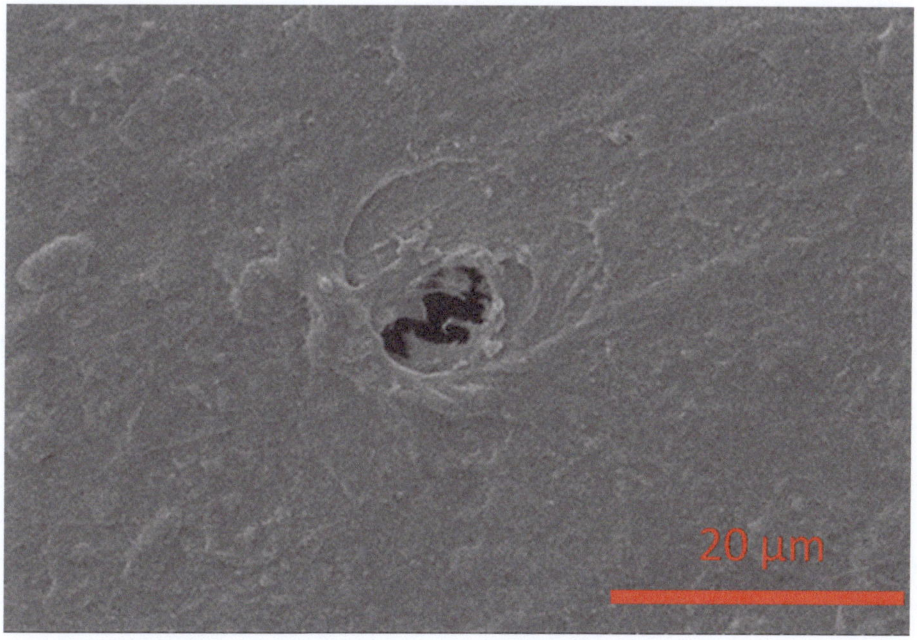

Fig. 3.12 Scanning electron microscope image of the RWM after punctuation with 20 μm tip

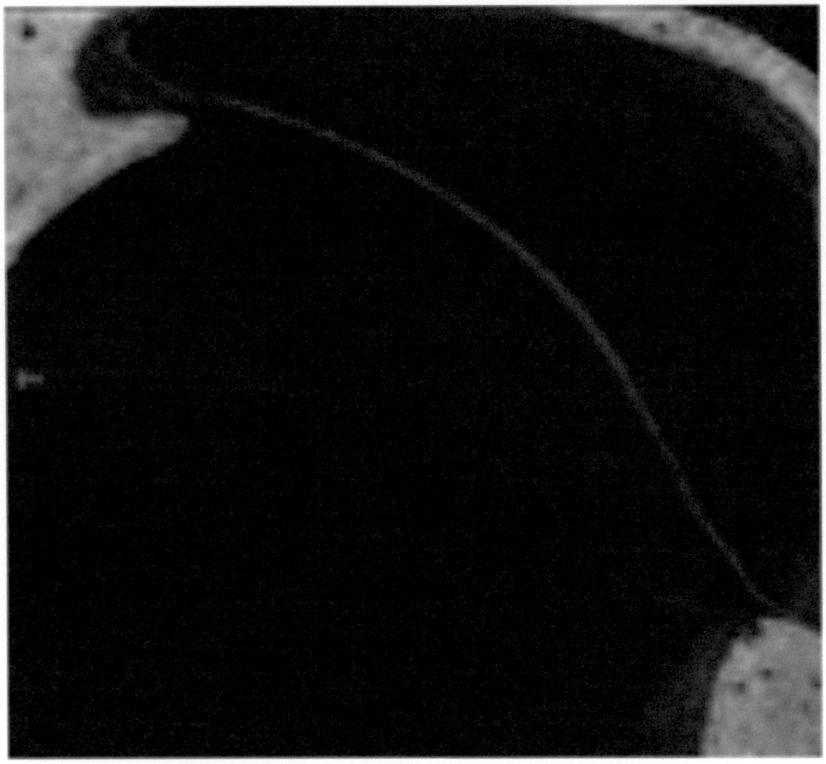

Fig. 3.13 A section of μCT scan showing soft tissue (*red*) and bone (*white*)

experimental curve. Figure 3.15 shows the predicted stress in radial direction under a probe tip. The red indicates that the rupture likely initiates around the center bottom of the RWM.

$$\sigma_m = \left(\frac{FE}{4\pi Rh} \right)^{\frac{1}{2}} \tag{3.1}$$

Equation (3.1) is derived by Bhatia et al. and gives the maximum stress σ_m, for a circular membrane with the thickness of h, clamped, linear elastic Young's modulus of E under a spherical indenter with the radius of R and applied load of F [23]. Using this equation, the maximum stress was calculated as 7.4 MPa.

3.4 Discussion

Sample Preparation of the RWM and Nanoindentation During the indentation, the RWM was housed in a humidity controlled space. Since the air should be saturated with humidity, dew condensation on the RWM was seen frequently. These water droplet on the RWM can affect the load-displacement curve, an additional experiment using a chamber that can submerge a RWM completely should be performed. The load-displacement curves tend to show some discontinuity suggesting either slips or multiple stages in the collagen fiber damage.

SEM Most notable and consistent feature of all of the perforations is the oval shape. Since the major axis of the oval was always horizontal to the major axis of the RWM in a plan view, these consistent results indicate the existence of anisotropy in the mechanical properties of the RWM. Overall, the size of the hole was readily controllable when a 20 μm probe tip was used. However, as the tip diameter decreased, the hole size became more difficult to control (data not shown). The hole size made with a 10 μm probe tip was more variable as well as the shape was distorted. With smaller tip down to sub-micron sized tip, the hole tended to be impossible to be found.

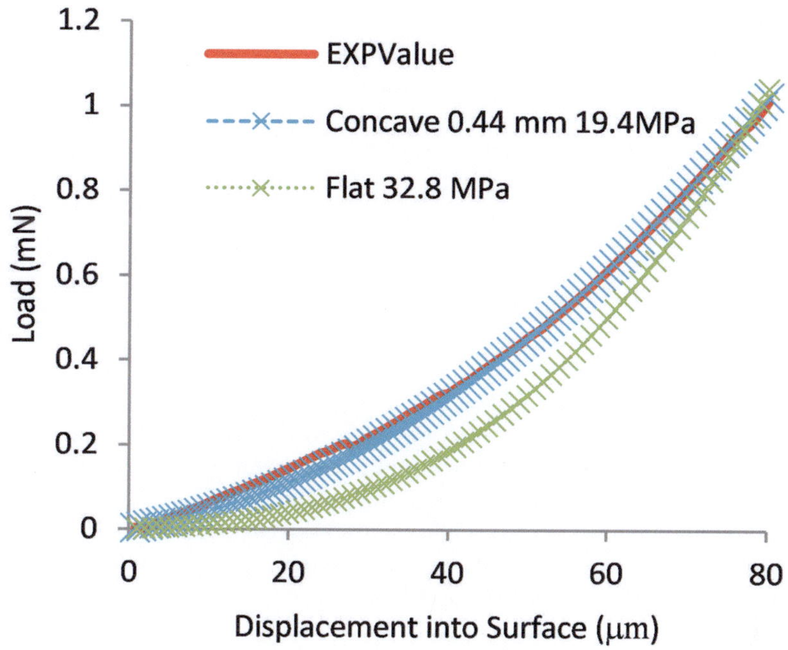

Fig. 3.14 Inverse analysis via ABAQUS to determine Young's modulus from Experimental data

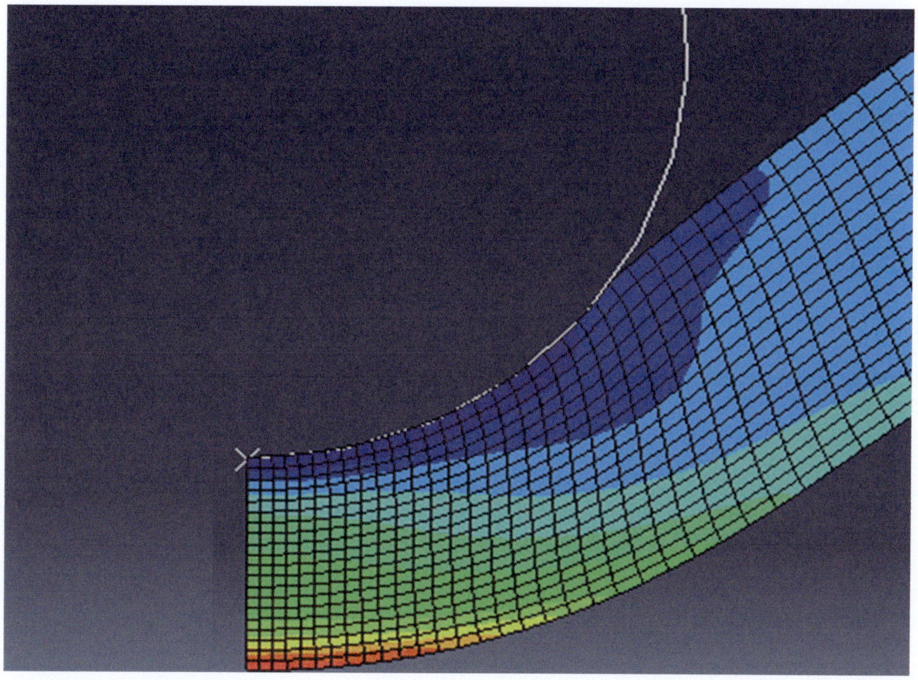

Fig. 3.15 ABAQUS simulation of a NANO indentation with axisymmetric model. The contour indicates the stress in radial direction

Micro CT In a μCT scan study, we need to be aware that the thickness in the scan may not be equal to the thickness of the actual membrane during the mechanical testing. The fixation process is not regarded as a significant factor that changes the shape. Depending on the threshold value set for delineating the hard and soft tissues, the μCT data tends to show slightly larger thickness than the value obtained with histological sectioning. Together with the possible thickness variation among the specimens, it would be ideal to have a data set as to the relationship between the thickness determined via μCT and that via histology as well as the μCT scan of each sample after indentation. Also, the μCT scan of each sample will provide the precise localization of the hole and boundary conditions.

Finite Element Modeling The results of inverse analysis suggested that it is necessary to include the contribution of the three dimensional effect from the saddle shape of the RWM. The Young's modulus calculated is larger than some of the literature values. First, the physical models some of the literature employed are likely to introduce significant errors due to the three dimensional effect as well as the lack of proper boundary conditions. It is well known that most of the biological tissues are under pretension and we have seen many evidence of the effect of pretension that expanded the holes into oval shapes consistently evidenced in the SEM micrographs. Since the collagen fibers are usually modeled as a hyperelastic material, lack of pretension in an experimental system may introduce decrease in the calculated value in Young's modulus.

3.5 Conclusion

In this study, an experimental approach to measurement of the static mechanical properties of RWM was developed. Nano-indentation using a micro-needle was performed to determine the load-displacement curve from the contact of the needle on the RWM specimen until the rupture caused by the needle indentation at the speed of 1 μm/s. The shape and the size of the membrane were confirmed using the data set obtained from previous study using micro CT. The experiments on RWM specimens were simulated in FE models using axis-symmetric model, and good agreement between the experimental measurements and FE modeling results was observed in indentation depth of 80 μm with the presence of concavity in the membrane. Static mechanical properties of the RWM were derived by the inverse-problem solving method. Young's modulus was 19.4 MPa and with curvature radius of 0.44 mm. Stress distribution suggests that the rupture of the membrane was initiated from the backside of the membrane by the tensile stress. Finally, this method of nano-indentation on the RWM may open a new approach to characterizing the bio-physical response of the RWM when the a micro-needle is used for the micro-perforation of the RWM to improve drug delivery into the inner ear.

Acknowledgments The authors thank Tizian Bucher, current graduate student in Mechanical Engineering Lab at the Columbia University for his technical assistance in preparing experimental set ups.. The authors also thank Elizabeth S. Olson Ph.D, at the Department of Biomedical engineering and Auditory Biophysics for generous support in animal studies, Elon J. Terrell Ph.D., at the Department of Mechanical Engineering at the Columbia University and Luis Cardoso Ph.D at the City University of New York. This work was supported by Coulter Translational Research Partnerships and American Otological Society grants.

References

1. Cheng, T., Gan, R.Z.: Experimental measurement and modeling analysis on mechanical properties of tensor tympani tendon. Med. Eng. Phys. **30**(3), 358–366 (2008)
2. Ishii, T., Takayama, M., Takahashi, Y.: Mechanical properties of human round window, basilar and Reissner's membranes. Acta Otolaryngol. Suppl. **519**, 78–82 (1995)
3. Mirza, S., Richardson, H.: Otic barotrauma from air travel. J. Laryngol. Otol. **119**(5), 366–370 (2005)
4. Gutovitz, S., et al.: Middle ear pressure and symptoms after skydiving. Aviat. Space Environ. Med. **79**(5), 533–536 (2008)
5. Yu, V.M., Strykowski, P.J., Odland, R.M.: A preliminary theoretical model of hydrodynamics in the inner ear. Ear Nose Throat J. **89**(4), 164–168 (2010)
6. Gan, R.Z., Nakmali, D., Zhang, X.: Dynamic properties of round window membrane in guinea pig otitis media model measured with electromagnetic stimulation. Hear. Res. **301**, 125–136 (2013)
7. Zhang, X., Gan, R.Z.: Dynamic properties of human round window membrane in auditory frequencies running head: dynamic properties of round window membrane. Med. Eng. Phys. **35**(3), 310–318 (2013)
8. Tringali, S., et al.: Intraoperative adjustments to optimize active middle ear implant performance. Acta Otolaryngol. **131**(1), 27–35 (2011)
9. Yoda, S., et al.: Round window membrane in Meniere's disease: a human temporal bone study. Otol. Neurotol. **32**(1), 147–151 (2011)
10. Yoon, Y.J., Hellstrom, S.: Ultrastructural characteristics of the round window membrane during pneumococcal otitis media in rat. J. Korean Med. Sci. **17**(2), 230–235 (2002)
11. Yoshida, M., Uemura, T.: Transmission of cerebrospinal fluid pressure changes to the inner ear and its effect on cochlear microphonics. Eur. Arch. Otorhinolaryngol. **248**(3), 139–143 (1991)
12. Aernouts, J., Aerts, J.R., Dirckx, J.J.: Mechanical properties of human tympanic membrane in the quasi-static regime from in situ point indentation measurements. Hear. Res. **290**(1–2), 45–54 (2012)
13. Aernouts, J., Soons, J.A., Dirckx, J.J.: Quantification of tympanic membrane elasticity parameters from in situ point indentation measurements: validation and preliminary study. Hear. Res. **263**(1–2), 177–182 (2010)
14. Daphalapurkar, N.P., et al.: Characterization of the linearly viscoelastic behavior of human tympanic membrane by nanoindentation. J. Mech. Behav. Biomed. Mater. **2**(1), 82–92 (2009)
15. Fay, J., et al.: Three approaches for estimating the elastic modulus of the tympanic membrane. J. Biomech. **38**(9), 1807–1815 (2005)

16. Volandri, G., et al.: Biomechanics of the tympanic membrane. J. Biomech. **44**(7), 1219–1236 (2011)
17. Zhang, X., Gan, R.Z.: Dynamic properties of human tympanic membrane based on frequency-temperature superposition. Ann. Biomed. Eng. **41**(1), 205–214 (2013)
18. Lim, K.M., Steele, C.R.: A three-dimensional nonlinear active cochlear model analyzed by the WKB-numeric method. Hear. Res. **170**(1–2), 190–205 (2002)
19. Kendall, M.A., Chong, Y.F., Cock, A.: The mechanical properties of the skin epidermis in relation to targeted gene and drug delivery. Biomaterials **28**(33), 4968–4977 (2007)
20. Decraemer, W.F., Dirckx, J.J.J., Funnell, W.R.J.: Shape and derived geometrical parameters of the adult, human tympanic membrane measured with a phase-shift moiré interferometer. Hear. Res. **51**(1), 107–121 (1991)
21. Ghiz, A.F., et al.: Quantitative anatomy of the round window and cochlear aqueduct in guinea pigs. Hear. Res. **162**(1–2), 105–112 (2001)
22. Watanabe, H., Kysar, J.W., Lalwani, A.K.: Microanatomic analysis of the round window membrane by white light interferometry and micro-computed tomography for mechanical amplification. Otol. Neurotol. **35**(4), 672–678 (2014)
23. Bhatia, N.M., Nachbar, W.: Finite indentation of an elastic membrane by a spherical indenter. Int. J. Non Linear Mech. **3**(3), 307–324 (1968)

Chapter 4
Transient Response of the Eardrum Excited by Localized Mechanical Forces

Payam Razavi, Ivo Dobrev, Michael E. Ravicz, Jeffery Tao Cheng, Cosme Furlong, and John J. Rosowski

Abstract The Tympanic Membrane (TM, eardrum) is the interface between the middle and outer ear and helps transform the variations in sound pressure in the ear canal into vibrations of the ossicles. However, the transient acoustic response of TM due to the complexity of wave interference, hinders the understanding of the motions. Therefore, to reduce this effect, local (<1 mm^2 area) mechanical excitation rather than entire-surface acoustic loading is proposed. To focus on the initial stages of the evolution of motion, we have developed a high-speed digital holographic system based on local phase correlation method, enabling the acquisition of full-field displacements of the TM up to 10 μs temporal and 150 k points spatial resolution. A piezoelectric plunger equipped with a feedback load sensor excites the TM by a 50 μs click and the mechanical contact is guaranteed during the experiment by >100 μm indentation of TM surface towards the camera. The results include characterization of wave travelling speed versus input force and localized mechanical properties, such as damping ratio, modal frequencies, and time constants. We expect that the results will lead to an improved understanding of the TM's localized material properties and modeling of the eardrum behavior.

Keywords Digital high-speed holography • Mechanical excitation • Otology • Transient response • Tympanic membrane

4.1 Introduction

Sound induced motions of the TM is the first stage of the hearing process, the TM transforms the sound energy into mechanical vibration in ossicles and causes the motion of hair cells in chochlea. Despite the past decades efforts of researchers, the complex motion of TM and its material properties have not yet been fully investigated [1–6]. Most of what is known of the transformation has been derived from measurements of the TM response to sound induced motion, however, most environmental sounds and speech contain acoustic transients such as click-like and chirp-like sounds. Based on our previous transient acoustic measurements, the motions start globally and are followed by complex motions that interfere, which limit the analysis and interpretation of their mechanics and extraction of mechanical properties [7, 11, 12, 13]. In this paper, we propose the use of our developed high speed digital holographic system and localized mechanical excitation to quantify the first stages of motion waves before their interference.

P. Razavi (✉)
Center for Holographic Studies and Laser micro-mechaTronics (CHSLT), Worcester, MA, USA

Mechanical Engineering Department, Worcester Polytechnic Institute, Worcester, MA, USA
e-mail: prazavi@wpi.edu

I. Dobrev
Center for Holographic Studies and Laser micro-mechaTronics (CHSLT), Worcester, MA, USA

University Hospital Zurich, Zurich, Switzerland

M.E. Ravicz • J.T. Cheng • J.J. Rosowski
Eaton-Peabody Laboratory, Massachusetts Eye and Ear Infirmary, Boston, MA, USA

Department of Otology and Laryngology, Harvard Medical School, Boston, MA, USA

C. Furlong
Center for Holographic Studies and Laser micro-mechaTronics (CHSLT), Worcester, MA, USA

Mechanical Engineering Department, Worcester Polytechnic Institute, Worcester, MA, USA

Eaton-Peabody Laboratory, Massachusetts Eye and Ear Infirmary, Boston, MA, USA

Department of Otology and Laryngology, Harvard Medical School, Boston, MA, USA

S.A. Tekalur et al. (eds.), *Mechanics of Biological Systems and Materials, Volume 6*,
Conference Proceedings of the Society for Experimental Mechanics Series, DOI 10.1007/978-3-319-21455-9_4

We also report progress of our efforts to improve the capabilities of previously developed High-speed Digital Holographic System (HDHS), which utilizes a novel $2+N$ frame acquisition method based on hybrid spatio-temporal Local Correlation (LC) phase sampling approach [7–9]. The HDHS has the advantage of utilizing the temporal resolution of a high-speed camera without limiting its spatial resolution and the need of specialized optical setups [7, 10]. Automatic execution and synchronization of high-speed measurements is achieved by a custom control system. The high temporal (i.e., > 40 kHz) and spatial (i.e., > 150 k data points) measurement resolutions of the HDHS enable the investigation of the complex spatio-temporal behavior of the TM at a sufficient level of detail to further expand our knowledge of the hearing processes.

4.2 Methods

We have developed and implemented a novel high-speed $2+N$ frame acquisition method based on hybrid spatio-temporal local correlation (LC) phase sampling approach [8, 10], that allows quantification of the TM transient deformations by utilizing two reference frames, $I_{ref}, I_{ref+\pi/2}$, and N consecutive deformed frames, $(I_{def})_{i,\, i\in 1,\, 2..N}$, recorded before and throughout the evolution of an event, as shown in Fig. 4.1.

According to Fig. 4.1b, the $2+N$ frame acquisition method starts by recording two phase $\pi/2$ phase shifted reference frames, *Iref* and $I_{ref+\pi/2}$, temporally separated to bypass the piezoelectric transducer (PZT) overshoot after the introduction of the phase shift [8]. While keeping the PZT at its $\pi/2$ position, a set of frames, $(I_{def})_{i,\, i\in 1,\, 2..N}$, i.e., deformed frames, are recorded at a rapid sampling rate (e.g., > 40 kHz) to capture the transient deformations of the TM. The corresponding spatio-temporal variation of the double-exposure optical phase change, $\phi(m,n,t)$, is quantified by correlating the intensities of each deformed and reference frames with

$$\phi(m,n,t) = tan^{-1}[\frac{\rho(I_{ref}(m,n),\, I_{def}(m,n,t))}{\rho(I_{ref+\pi/2}(m,n), I_{def}(m,n,t))}], \tag{4.1}$$

where ρ is the Pearson's correlation coefficient for finite discrete sets having the recoded intensities as arguments and computed based on a spatial kernel (i.e., with dimensions of 3×3 or 5×5 pixels) centered around each measurement point, (m,n), of a pair of reference and deformed frames [7, 8].

There is only one phase shift and the LC method is highly sensitive to the exact degree of the shifted phase. Furthermore, due to the physical constraints of the phase shifter device, as shown in Fig. 4.1b, the phase shifter experiences an overshoot and ringing after the $\pi/2$ phase shift. Additionally, there are other environmental factors such as room temperature and driver resolution which cause random inaccuracies in the phase shifter. To overcome the sensitivity of the LC method, we propose a continuous rather than a sudden phase shift in order to minimize the effect of position overshooting and randomness, as shown in Fig. 4.1c. In addition, by using a phase quality optimization algorithm, the frames corresponding to the best phase quality are chosen. This modification in the acquisition algorithm, has resulted in more repeatable and stable HDHS results. It has also eliminated the precision issues caused by nanometer errors introduced by the phase shifter and/or movements of other elements in the setup.

4.3 Representative Results

A post mortem human sample was characterized with the updated HDHS setup and with mechanical and acoustic stimuli. The TM was mechanically excited from the ossicular chain side and the middle ear was detached from the inner ear leaving some of the stabilizing ligaments and stapes intact. In this preparation, the sample maintains nearly all of its natural behavior during local excitation. A custom PZT actuator mounted on a three dimensional positioner excited the membrane locally at different regions, as shown in Fig. 4.1a. Due to the advantage of high-speed data acquisition, multiple sets of data can be acquired with negligible environmental effects, such that boundary condition effects, including changes in the sensitivity vector, sample dehydration, and excitation angle are negligible and the results are precise. Typical measurements consisting of at least 30 mechanical excitations at multiple locations and 12 acoustic excitations last less than 2 h. Representative results of individual acoustical and mechanical excitation measurements are presented.

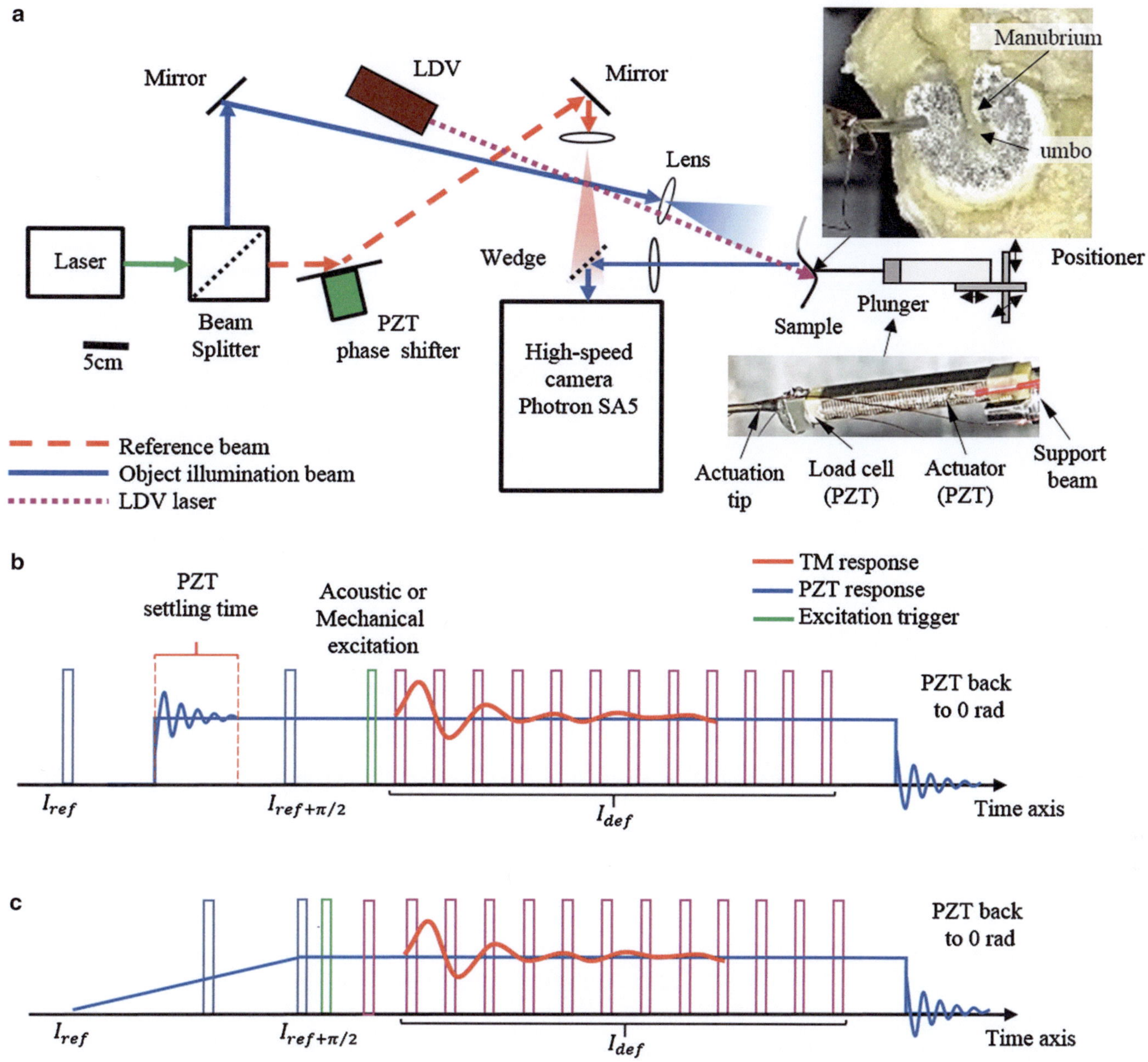

Fig. 4.1 Schematic representation of the HDHS and the acquisition timing diagrams: (**a**) HDHS setup showing the human TM sample subjected to localized excitation with a custom made plunger; (**b**) former timing diagram demonstrating the principle of operation of the high-speed 2+N frame acquisition method [7]; and (**c**) the modified continuous phase shifting method that reduces the effects of PZT oscillations at the beginning of data acquisition

4.3.1 Acoustic Excitation

The natural responses of the TM subjected to loading by acoustic clicks having a typical duration of 50 μs were investigated. The HDHS was used to measure the transient time waveform response of the TM at each point across the full-field-of-view, as shown in Fig. 4.2. Figure 4.2b depicts the deformations of the TM at six specific points of interest, including the umbo. Analysis of the deformations of the entire TM indicates that the maximum deformation occurs at 595 μs from the beginning of the acoustic excitation. At this state the manubrium and umbo experience their maximum deformations, which are followed by oscillations of other parts of the TM that include travelling waves as well as wave interference while the oscillations of the umbo are dampened rapidly. At the later stages of oscillations, the interpretation of the mechanics is challenging, typically 600–900 μs after the umbo reaches its maximum deformation.

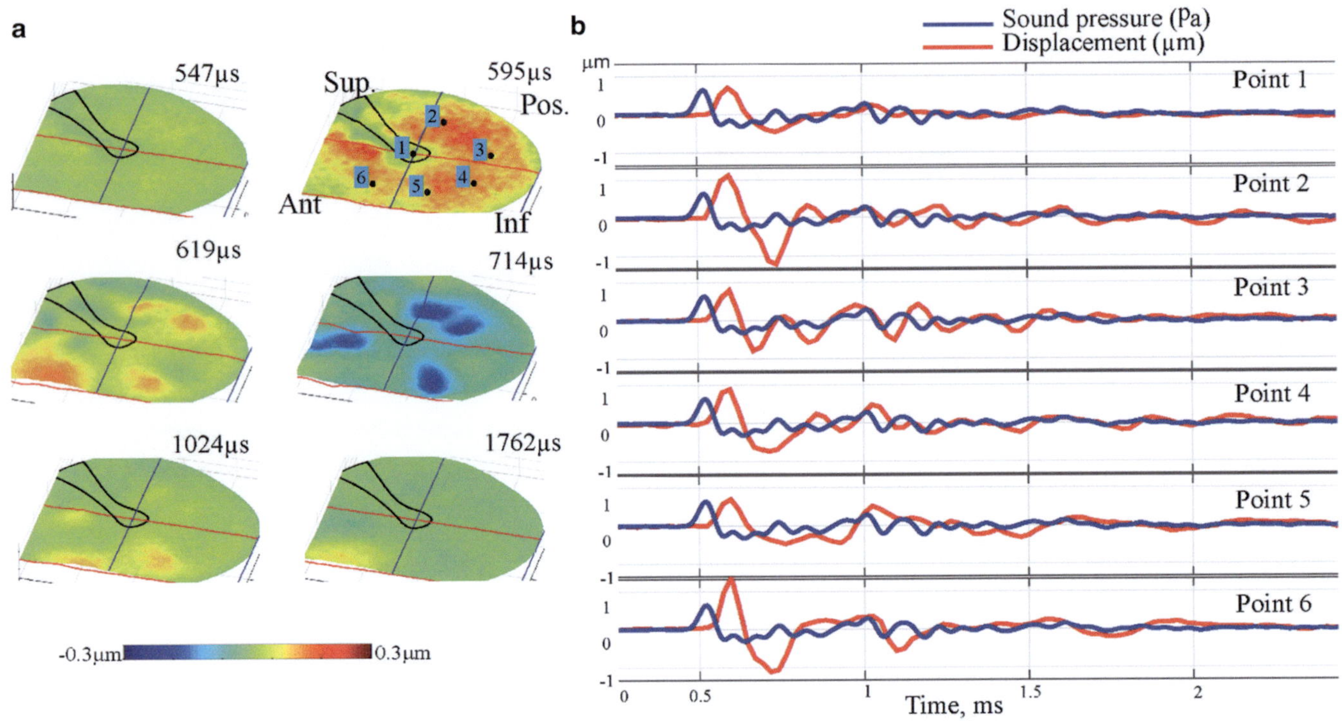

Fig. 4.2 Representative transient measurements of the TM subjected to acoustic excitation: (**a**) selected full-field motions at six temporal instances spanning the period from the beginning of the excitation to 1.7 ms; and (**b**) deformation and sound pressure vector at six specific points of interest as identified in (**a**) at 595 μs

By analyzing the transient response of each 147 k points due to the impulse excitation, the natural frequencies and mode shapes up to 12 kHz can be calculated, including corresponding time constants, acoustic phase delay, mechanical damping, and local material properties. These characteristics are determined by analyzing the envelope of the decaying function in the time-displacement domain for every single point, as shown in Fig. 4.3.

4.3.2 Mechanical Excitation at Umbo

The role of the human tympanic membrane is to transform sound into mechanical energy and the umbo links the TM to the ossicular chain. Therefore, exciting the TM at the umbo location can help understand the energy transformation from the TM to the umbo and vice versa.

Figure 4.4 shows representative full-field measurements of transient deformations of the TM when mechanical excitation is applied at the umbo. The excitation was a 50 μs square wave with amplitude of 500 mV and in order to maintain contact between the plunger and the TM during the measurements, the plunger was preloaded about 50 μm into the surface of the TM. The results show that uniform deformation waves begin at the umbo and propagate around the manubrium to reach the TM ring. While the manubrium settles down after around 100 μs the rest of the TM continues to oscillate for about 1.2 ms.

4.3.3 Mechanical Excitation at Other Regions

Following the mechanical excitation of the umbo, the TM was mechanically excited at other locations. To enable these measurements, the plunger was mounted on a three dimensional positioner to locate it at the following specific points of interest on the membrane, as shown in Fig. 4.5,

a) posterior superior,
b) posterior inferior,
c) anterior inferior.

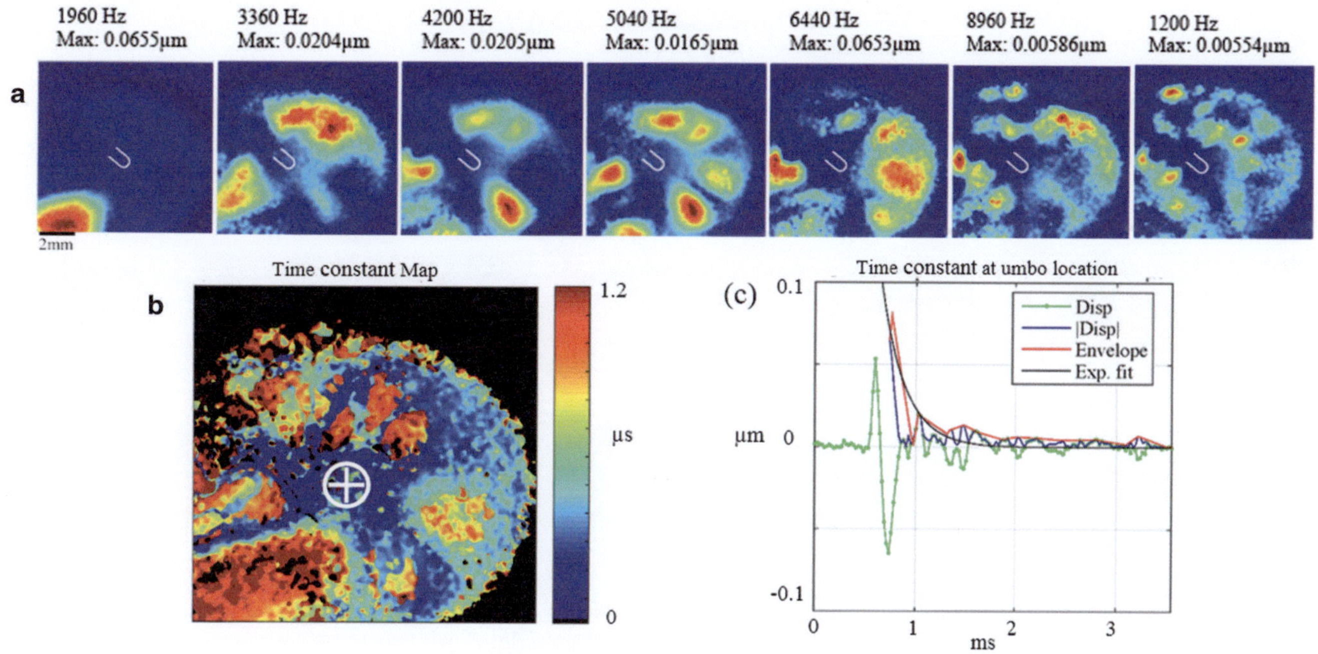

Fig. 4.3 Quantification of the acoustic stimuli results: (**a**) corresponding mode shapes, frequencies, and corresponding maximum displacements; (**b**) The time constant map of the full spatio-temporal motion, and; (**c**) time constant waveform at the umbo location marked by *circled plus* at part (**b**)

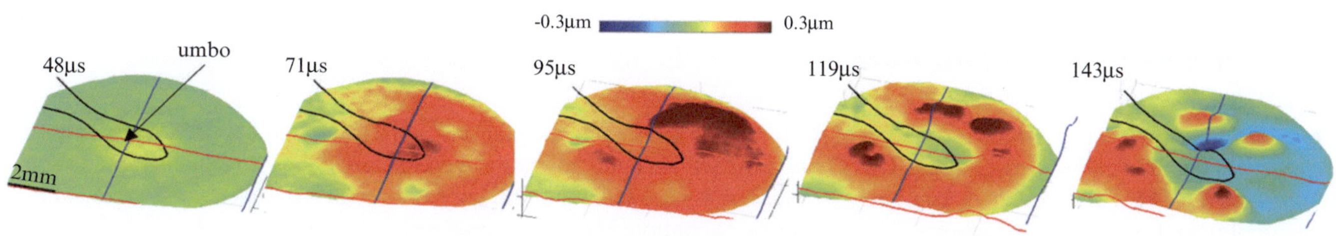

Fig. 4.4 Full-field of view displacement of the mechanical excitation at the umbo location

Figure 4.5 shows the mechanical response of the membrane from 71 to 167 μs, relative to the beginning of the excitation, for the three points of interest. We observed that after about 200 μs, the complexity of the motions increase to the point that their analysis and interpretation become cumbersome. Although we plan to investigate such complex motions in future works, some preliminary observations can be extracted from our current measurements. Specifically, for location (a), Fig. 4.3a, measurements indicate an overall higher amplitude of motions than at the other locations, (b) and (c), by about 0.2 μm. The motion starts clockwise with respect to the umbo as the center of rotation. We assume that the higher deformation is partially caused by a bony part of the manubrium located at the top left of the excitation, which stops the waves from moving counterclockwise. In this case all of the waves travel clockwise with a faster speed than at other locations. The arrows shown in Fig. 4.5 indicate the direction of wave propagation. The motion starts with a speed around 60–90 m/s and reaches the other side of the TM along the radial direction and it is followed by waves hitting the tympanic ring at the superior anterior quadrant and returning to the location of excitation. For locations (b) and (c), Fig. 4.5b–c, the motion spreads in both, clockwise and counterclockwise directions and the measured traveling wave speeds are around 40 m/s for location (b) and as low as 20 m/s for location (c). The maximum amplitude of displacement shows a reduction from location (a) to location (c). In all three locations, the manubrium undergoes reduced motions. The waves produced by mechanical excitation at the *pars tensa* of the TM tend to dissipate by interference rather than moving the umbo, which suggests that the TM dissipates local motions and filters them out. These measurements also suggest spatial variations of damping characteristics and mechanical properties across the TM surface. The shape of the deformations produced at the first moments of the mechanical impact as well as their temporal evolution can reveal the type of material model, e.g., isotropic or orthotropic, characterizing

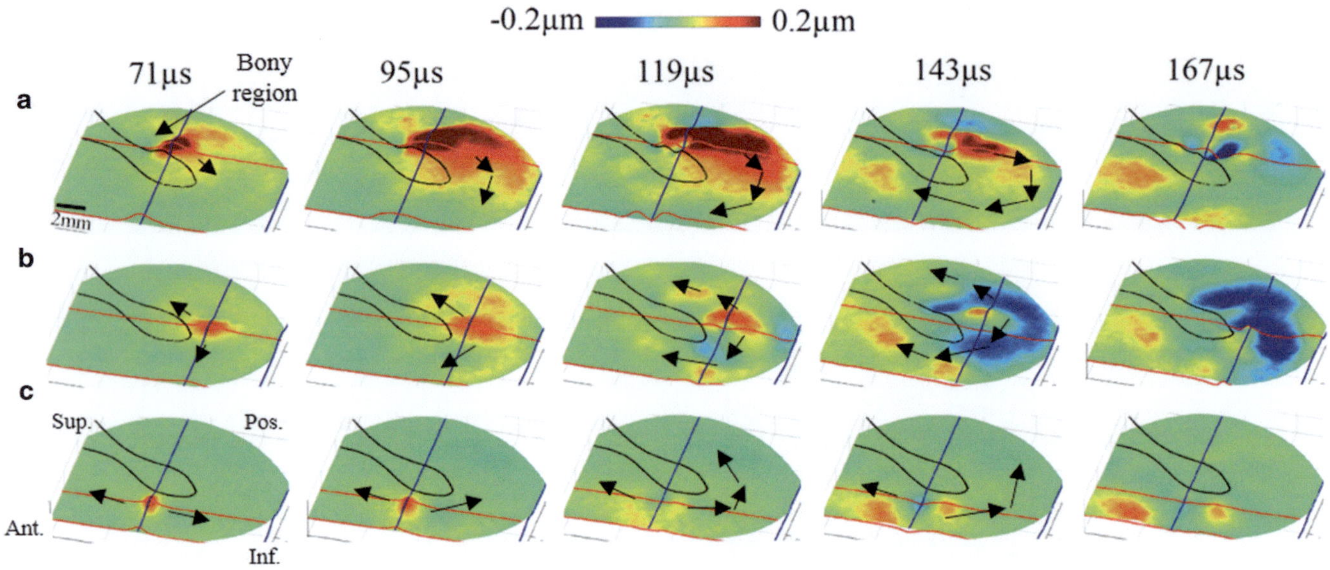

Fig. 4.5 Motion of the TM due to mechanical excitation at the three stated positions: (**a**) posterior anterior; (**b**) inferior posterior, and; (**c**) superior anterior (*Sup.* stands for superior, *Pos.* stands for posterior, *Inf.* Stands for inferior and *Ant.* stands for anterior)

the material properties of the TM. Based on the measurements at locations (a) to (c) different wave propagation patterns and wave speeds can also suggest spatial variations of the thickness of the TM.

4.4 Conclusions and Future Work

Transient acoustic and mechanical excitation of the tympanic membrane are potential approaches to characterize the acousto-mechanical mechanical properties of the TM. There are a few available methods to characterize the transient nanometer scale motions of the TM with a sufficient temporal resolution throughout the audible range and with a spatial resolution to capture the complex motions that are developed during acoustic excitation. We have been developing high-speed digital holographic systems to measure transient events in the middle ear [7, 8]. In this paper, we report improvements to our novel HDHS system to further enhance its measuring accuracy and precision. By applying the updated HDHS to investigate acoustic (global) as well as mechanical (local) stimuli of the TM, we are able to capture the complex transient motions that are developed. Acoustic and mechanical transients provide distinct and complementary stimuli for the study of TM response. Further investigations includes analysis of the capabilities and the effectiveness of the two stimuli methods to determine TM properties such as spatial variations in dominant frequency, time constant, and wave propagation speeds. HDHS is enabling fundamental research on the initial transient dynamics of the TM, its relationship to the energy transfer into the middle-ear, and its connection to previous steady-state dynamics. HDHS will be further optimized for in-vivo applications in medical research and results will be reported in future publications.

Acknowledgements This work has been funded by the National Institute on Deafness and Other Communication Disorders (NIDCD), the National Institute of Health (NIH), the Massachusetts Eye and Ear Infirmary (MEEI), and the Mittal Fund. The authors should like to acknowledge the help of Morteza Khaleghi at the Center for Holographic Studies and Laser micro-mechaTronics (CHSLT) at Worcester Polytechnic Institute.

References

1. Rosowski, J.J., Cheng, J.T., Ravicz, M.E., Hulli, N., Hernandez-Montes, M., Harrington, E., Furlong, C.: Computer-assisted time-averaged holograms of the motion of the surface of the mammalian tympanic membrane with sound stimuli of 0.4–25 kHz. Hear. Res. **253**(1), 83–96 (2009)
2. Rosowski, J.J., Dobrev, I., Khaleghi, M., Lu, W., Cheng, J.T., Harrington, E., Furlong, C.: Measurements of three-dimensional shape and sound-induced motion of the chinchilla tympanic membrane. Hear. Res. **301**, 44–52 (2012). doi:10.1016/j.heares.2012.11.022

3. Khaleghi, M., Furlong, C., Ravicz, M., Cheng, J.T., Rosowski, J.J.: Three-dimensional vibrometry of the human eardrum with stroboscopic lensless digital holography. J. Biomed. Opt. **20**(5), 051028 (2015)
4. Puria, S., Allen, J.B.: Measurements and model of the cat middle ear: evidence of tympanic membrane acoustic delay. JASA **104**, 3463 (1998)
5. Tonndorf, J., Khanna, S.M.: Tympanic-membrane vibrations in human cadaver ears studied by time-averaged holography. J. Acoust. Soc. Am. **52**, 1221–1233 (1972)
6. Cheng, J.T., Hamade, M., Merchant, S.N., Rosowski, J.J., Harrington, E., Furlong, C.: Wave motion on the surface of the human tympanic membrane: Holographic measurement and modeling analysis. JASA **133**, 918 (2013)
7. Dobrev, I., Furlong, C., Rosowski, J.J., Cheng, J.T.: High-Speed Digital Holography for Transient Response of the Human Tympanic Membrane. Adv. Opt. Methods Exp. Mech. **3**, 337–342 (2015). doi:10.1007/978-3-319-06986-9_39
8. Dobrev, I., Furlong, C., Rosowski, J.J., Cheng, J.T., Harrington, E.J.: Implementation and evaluation of single frame recording techniques for holographic measurements of the tympanic membrane in-vivo. Proc. SEM **2013**(3), 85–95 (2014)
9. Georgas, P.J., Schajer, G.S.: Modulo-2pi phase determination from individual ESPI images. Opt. Lasers Eng. **50**(8), 1030–1035 (2012)
10. Pedrini, G., Osten, W., Gusev, M.E.: High-speed digital holographic interferometry for vibration measurement. Appl. Opt. **45**(15), 3456–3462 (2006)
11. Kemp, D.T.: Stimulated acoustic emissions from within the human auditory system. JASA **64**, 1386 (1978)
12. Puria, S.: Measurements of human middle ear forward and reverse acoustics: implications for otoacoustic emissions. JASA **113**, 2773 (2003)
13. Decraemer, W.F., Khanna, S.M., and Funnell, W.R.J.: Vibrations at a fine grid of points on the cat tympanic membrane measured with a heterodyne interferometer. In: EOS/SPIE International Symposia on Industrial Lasers and Inspection, Conference on Biomedical Laser and Metrology and Applications, 1999

Chapter 5
In Vitro Quantification of Optimal Impact Properties for Microneedle Penetration

Kikelomo Moronkeji, Simon Todd, Idalia Dawidowska, and Riaz Akhtar

Abstract Microneedles provide a minimally invasive means of delivering drugs into the body via the skin. The challenge to effective microneedle penetration is to bypass the inherent elasticity of the stratum corneum, the skin's uppermost layer. However, there are few studies in the literature which focus on optimising parameters for repeatable penetration into skin. Here, a dropped weight impact testing rig was developed to allow high-density polymethylmethacrylate (PMMA) microneedles to be applied on to neonatal porcine skin at various velocities and forces. Methylene blue and histological staining were used to determine penetration effectiveness. Whilst methylene blue staining confirmed successful penetration of the stratum corneum, subsequent histological imaging demonstrated how deeply the microneedles had breached the layers of skin. The optimal parameters of velocity and force for reproducible and predictable penetration of the skin were found to be 4.52–4.8 ms^{-1} (nominal velocity of 3 ms^{-1}) and 3.35–4.17 N.

Keywords Microneedles • Impact testing • Porcine skin • Stratum corneum • Penetration

5.1 Introduction

Skin is a multi-layered tissue that acts as the main organ of protection against the external environment. A vital role of skin is the protection against external mechanical assaults, which is achieved by a reversible deformation of its structure [1].

The skin is composed primarily of three layers, which are the stratum corneum (SC), epidermis, and dermis [2]. The SC, which is several layers of flat, keratinized, nonnucleated, dead cells, serves to protect the internal organs and prevents foreign matter from permeating into the body. By overcoming the normal, protective function of the SC, the skin offers the potential for multiple sites which can be accessed for the administration of therapeutic compounds. Over the years, several methods and technologies have been developed to bypass the SC layer, such as jet injectors [3], thermal poration [4], iontophoresis [5], laser [6] and the transdermal microneedle device [7].

Microneedle arrays (MNs) are a promising minimally invasive means of delivering drugs into the body or extracting interstitial fluids, for example from the skin for glucose measurement [8]. They are submillimeter projections capable of penetrating the SC and are typically made from a range of different materials including silicon [9], polysilicon [10], polymers [11], metal [12] and sugars [13]. MNs are typically fabricated as an array of up to several hundreds of needles over a base substrate. The MNs can be fabricated as solid arrays, solid arrays coated with therapeutic compound, biodegradable arrays of MNs and hollow MNs [14]. The MNs are designed to pierce the SC and deliver the therapeutic compound into to the epidermal layer. As there are no nerve endings or vasculature in the epidermis, MN drug delivery is not associated with pain and bleeding as compared to conventional hypodermic needles [14], and the micron-scale holes in skin created by MNs provide a user-friendly means of drug delivery [15].

The extent to which MNs increase skin permeability is related to their design, with consideration given to parameters such as tip radius, base radius, pitch, number of MNs per array and penetration depth [16], as well as the mechanical behaviour of skin during its interaction with MNs. Previous literature details the study of skin penetration mechanics using hypodermic needles [17], knives [18] and punches [19]; however, there is little work on skin penetration using MNs. This is despite the

K. Moronkeji • R. Akhtar (✉)
Centre for Materials and Structures, School of Engineering, University of Liverpool, Liverpool L69 3GH, UK
e-mail: r.akhtar@liverpool.ac.uk

S. Todd • I. Dawidowska
Renephra Ltd, MedTech Centre, Manchester Science Park, Pencroft Way, Manchester M15 6JJ, UK

S.A. Tekalur et al. (eds.), *Mechanics of Biological Systems and Materials, Volume 6*,
Conference Proceedings of the Society for Experimental Mechanics Series, DOI 10.1007/978-3-319-21455-9_5

wide study of microneedles for pharmaceutical applications over the past 10 years [20]. Donnelly et al. [21] demonstrated that a large insertion force applied at low speed barely penetrated the SC, however, a smaller force at a higher speed pierced it. Verbaan et al. [22] demonstrated that with an electrical applicator MN penetration could be improved with a higher piercing velocity. Their setup provided the ability to deliver low density arrays MNs into dermatomed human skin at 1 or 3 ms^{-1}. More recently, van der Maaden et al. [23] demonstrated that the SC layer could be overcome in a reproducible manner using an impact-insertion applicator at a fixed velocity of 3 ms^{-1} and forces ranging from 3.43 to 22.1 N.

The motivation of this paper was to quantitatively determine the parameters required for repeatable and reproducible insertion of high-density MNs into skin using a dropped weight impact test rig. A representative physical model of skin that includes the underlying layers such as muscle was constructed. The penetration efficiency with this setup was assessed using methylene blue (MB) staining and histology.

5.2 Materials and Methods

5.2.1 Materials

A fresh suckling pig (7–10 weeks old) was obtained from a local abattoir. Skin samples from the back were dissected within 3–4 h and immediately snap-frozen by placing in super-cooled isopentane, and were subsequently stored at −80 °C until required. All the chemicals used in this study were obtained from Sigma-Aldrich (Dorset, UK). Commercial hair removal cream for normal skin (Veet, Reckitt Benckiser Group, Berkshire, UK) was applied to the skin surface for 5 min to remove any hair prior to testing. Perma-Gel was obtained from a third party distributor (MidwayUSA, Columbia, USA).

Dense Polymethylmethacrylate (PMMA) MN arrays were used for all experiments (10x technology, IL, USA). The full specifications for these MNs are presented Table 5.1.

5.2.2 In Vitro Skin Model

A three layer *in vitro* skin model was constructed composing of neonatal porcine skin, gelatine gel (88 % water content) and Perma-gel as shown in Fig. 5.1. Neonatal porcine skin was used due to its similar mechanical and physical properties to human skin as well as its lack of hair as compared to mature porcine skin [24]. The high water content gelatine layer acted as a representative layer for subcutaneous tissue. Perma-gel, a synthetic thermoplastic material is similar to ballistic gelatin in its properties [25] was used as a representative layer for muscle tissue.

Prior to testing, the skin tissue was de-thawed for 30 min. Subsequently, hair was removed from the skin and the samples were kept hydrated in phosphate buffered saline (PBS). Gelatine gel of 88 % water content by weight was prepared and cured at −20 °C. The Perma-Gel was cut and fitted to custom mechanical clamps (Fig. 5.1).

5.2.3 In Vitro Impact Microneedle Applicator

A dropped weight impact-testing rig was constructed in order to apply MNs at penetration speeds in excess of 2 ms^{-1}, in line with other research on microneedle impact [23]. The experimental setup is shown in Fig. 5.2.

The dropped weight impact testing rig comprised a drop tower of height 1.86 m, with a drop tower guide of height 1.78 m, a piezoelectric accelerometer with 10.71 mV/g sensitivity (Model 3225F-1, Dytran Instruments, Inc., CA, USA) and a

Table 5.1 PMMA MNs specification

Length	Disc size (diameter, mm)	Active area (mm^2)	Needles per unit area (mm^2)	Height (μm)	Base (μm)	Centre-to-centre spacing (μm)	No. of needles per array
356 nm	15.8	196	6.7	356	177	386	1316

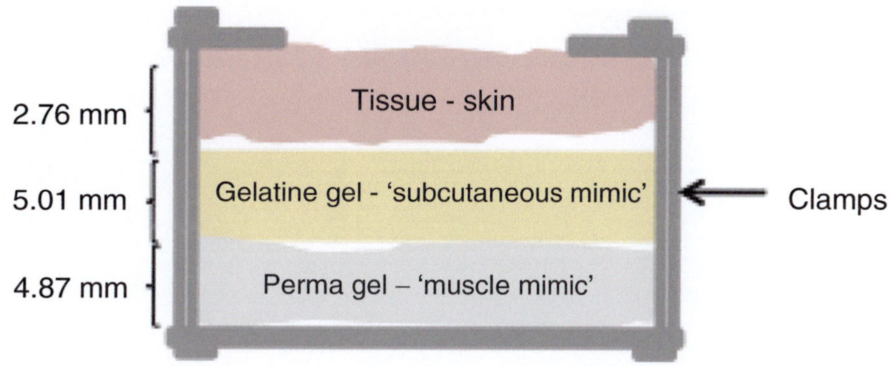

Fig. 5.1 In vitro skin model

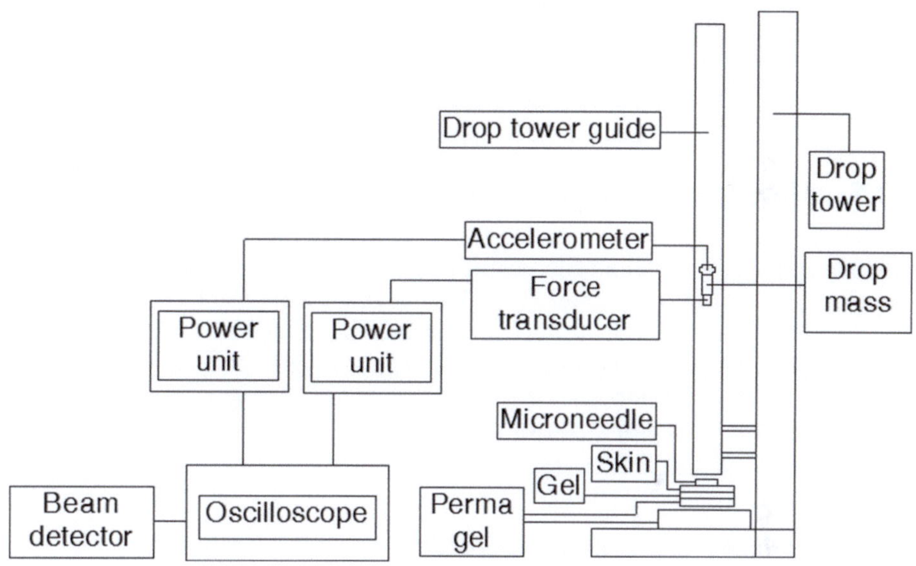

Fig. 5.2 Schematic diagram of the dropped weight impact-testing rig

piezoelectric force transducer with 22.4 mV/N sensitivity (Model 1022V, Dytran Instruments, Inc., CA, USA). Both the accelerometer and the transducer were connected to their respective amplifiers and to a digital oscilloscope (54641A, Keysight Technologies, Berkshire, UK) and utilized with Keysight IntuiLink software (Version 3.1). The accelerometer was connected to the top of the custom-made drop mass, whilst the force transducer was fastened to its bottom end. Two 630–650 nm wavelength (red) with a power output <1 mW laser diode modules (PL-Series, Hero Electronics, Dunstable, UK) were attached to two drill holes on the lower part of the drop tower guide so that a break in signal, caused when the drop mass is released, is captured by a beam detector, thus activating the external trigger on the oscilloscope. Subsequently, the force and acceleration outputs were captured on the oscilloscope.

5.2.4 Calibration of the In Vitro Impact Applicator

The dropped weight impact testing rig was calibrated using three aluminium drop masses weighing 7, 17 and 36 g. The drop mass was released multiple times at each dropping height onto a Perma-Gel block. The acceleration and the drop-mass force were obtained from the oscilloscope as voltage–time plots. The voltage–time plots were converted to acceleration–time and force–time plots using the sensitivities of the accelerometer and force transducer respectively. The data from the acceleration–time plot was multiplied by time and converted to a velocity–time plot.

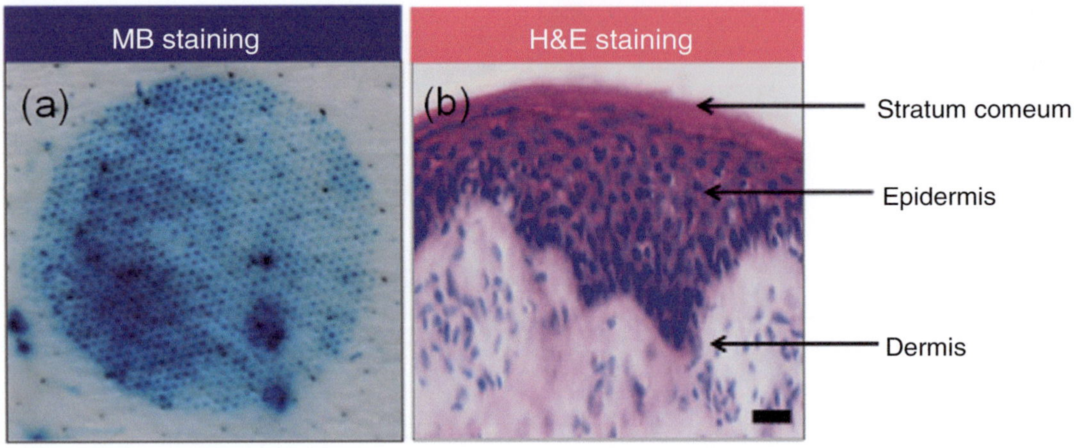

Fig. 5.3 Inspection technique of the perforated tissue. (**a**) MB staining of the skin. Diameter of the MN disc is 15.8 mm. (**b**) H&E staining of the sectioned tissue with the three histological layers of skin visible. Scale bar represents 100 μm

5.2.5 Microneedle Penetration Experiments with the In Vitro Applicator

The impact applicator was initially tested with the microneedles placed on the Perma-Gel, at five different release heights along the length of the drop tower guide. Each release height had its nominal velocity equivalent, ranging from 1 to 5 ms^{-1}. The tests were repeated with the three drop masses (7, 17, 36 g).

Subsequent testing was conducted by placing the microneedle array on the surface of the skin model (Fig. 5.1) with a Perma-Gel block at the bottom of the drop tower guide. The same drop release heights were used and the tests were again repeated with the three drop masses.

5.2.6 Imaging of Penetration Quality

Following completion of the impact testing, the penetration 'quality' was assessed with methylene blue (MB) solution and haematoxylin & eosin (H&E) staining. MB staining indicates breach of the stratum corneum and creation of microchannels (Fig. 5.3a) within the skin [26]. H&E provides conventional histological images of skin (Fig. 5.3b) and is useful for assessing the depth of microchannels created within the skin. For MB imaging, the skin was immediately soaked in MB solution and the skin samples were imaged with Nikon digital camera D5100 (Surrey, UK). For H&E imaging, the skin was fixed in 4 % paraformaldehyde and stained with H&E. The stained tissue was then sectioned to 12 μm thickness using a Leica CM1850 cryostat (Milton Keynes, UK). The sections were imaged with a Nikon Eclipse Ci microscope (Surrey, UK).

5.3 Results and Discussion

5.3.1 Calibration Tests of In Vitro Impact Test Set-Up

Figure 5.4 shows the linear relationship between force and velocity. Due to the contribution from impact acceleration, however, the force output was found to be more than 50 times greater than the drop mass load. This trend was consistent on all force–time plots.

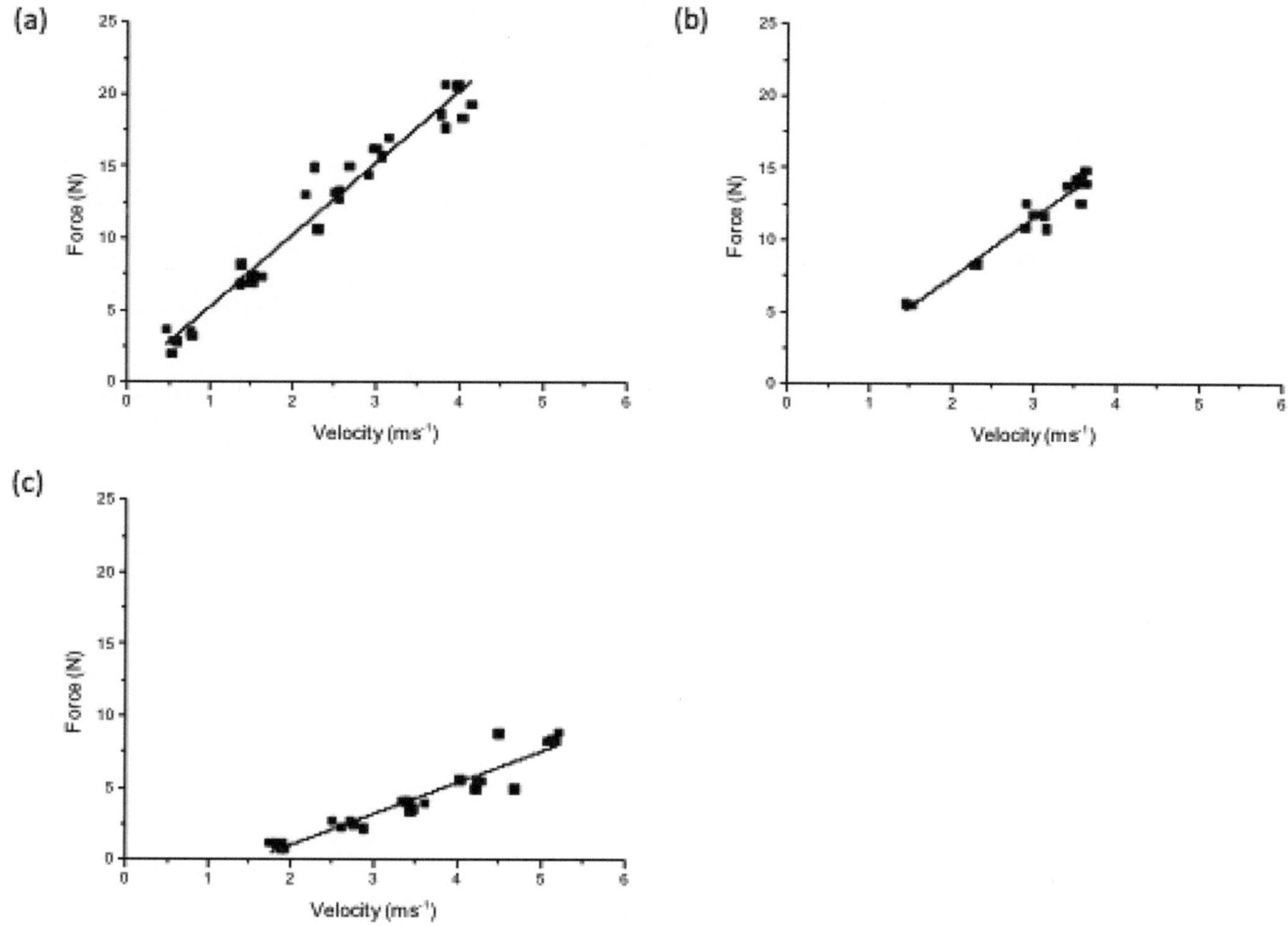

Fig. 5.4 Linear plots of velocity as a function of force for calibration tests on Perma-Gel using various drop masses weighing. (**a**) 36 g, (**b**) 17 g and (**c**) 7 g

5.3.2 Perma-Gel and Microneedle Interaction

The linear relationship between force and velocity was also found when the microneedles were impacted into the Perma-Gel block. The trends for the three different drop masses are shown in Fig. 5.5. A linear relationship between velocity and force for each of the three drop masses was observed, similar to the trend found with the calibration tests. The velocity and force outputs were much higher as compared to the calibration tests.

5.3.3 Microneedle Penetration of Skin Following Impact

Figure 5.6a–d show the velocity–time and force–time plots at various nominal velocities (i.e. determined from the different drop heights). The values of force and velocity were calculated from the peak of the curves for each plot, as illustrated in Fig. 5.5e. The calculated velocities ranged from 4.52 to 7.56 ms^{-1}, which is equivalent to nominal velocities of 2–5 ms^{-1}. As seen in Fig. 5.6a–d, there was a second peak on the curves, which was due to the rebound from the drop mass on impact, as the residual energy was dissipated. For all the plots, the force outputs ranged from 2.15 to 9.75 N.

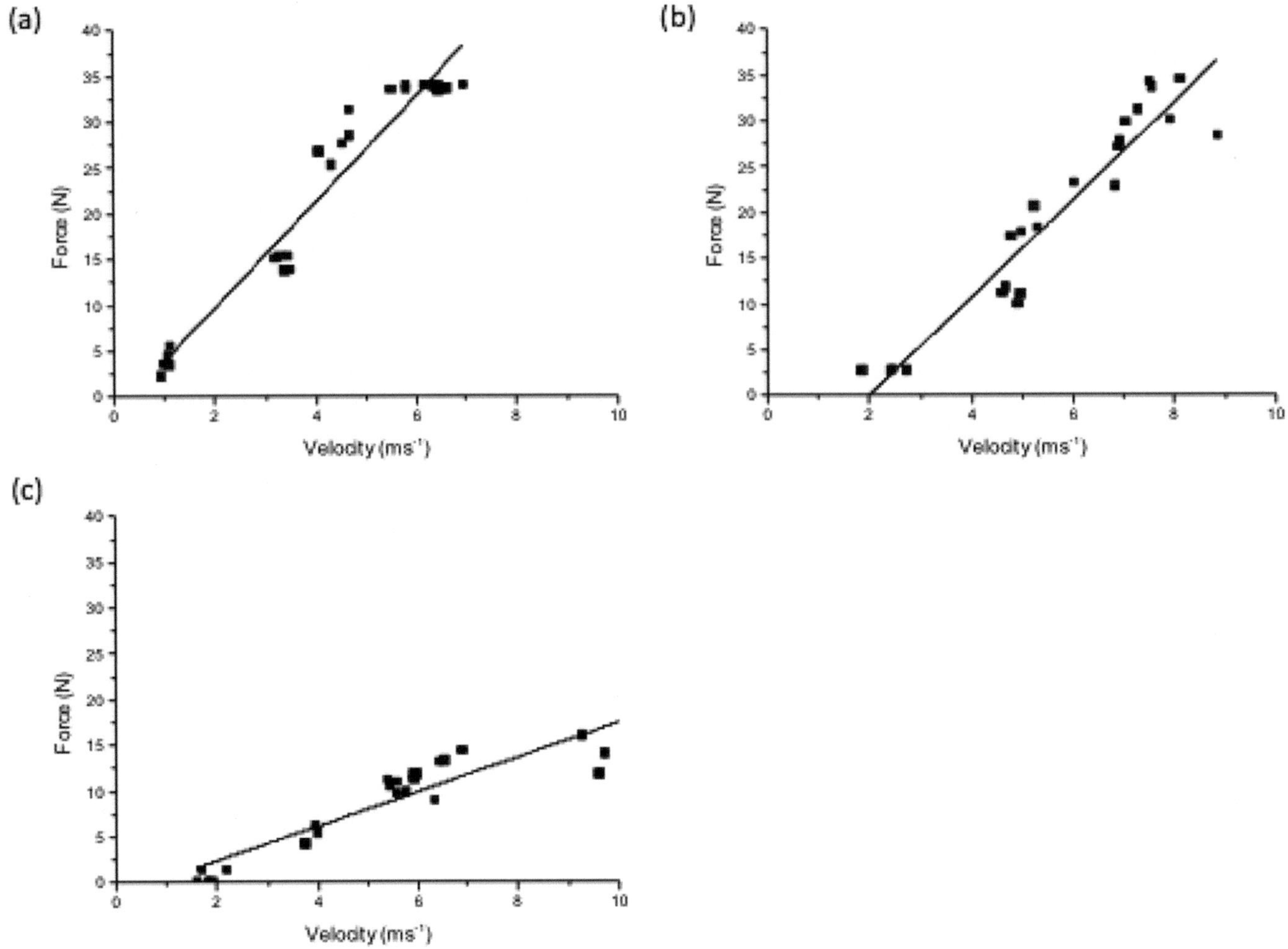

Fig. 5.5 Linear plots of velocity as a function of force for impact tests of MNs resting on Perma-Gel using various drop masses weighing. (**a**) 36 g, (**b**) 17 g and (**c**) 7 g

Penetration of the stratum corneum was assessed qualitatively from the MB images as shown in Fig. 5.7. Significant MB staining was observed with both the 36 and 17 g drop mass at nominal velocities of 5 and 4 ms^{-1}. At 3 ms^{-1}, a significant proportion of the MNs penetrated the SC with the 36 g drop mass (Fig. 5.7a) whereas penetration appeared substantially lower with the 17 g drop mass (Fig. 5.7b). At 2 ms^{-1}, the intensity of MB staining was further reduced with the 36 g drop mass (Fig. 5.7a) but minimal staining was observed at this velocity with the 17 g drop mass (Fig. 5.7b). These images demonstrate that the MN arrays successfully perforated the skin at velocities ≥ 3 ms^{-1} with the highest drop mass used in this study. The quality of penetration decreased with the lower drop mass weights.

Microneedles are designed not only to penetrate the stratum corneum but to then effectively deliver drugs into the epidermal layer without causing any damage to the dermal layer. Hence, the histology images were used to examine the penetration profile alongside the MB images. In some cases, only minute breaches in the epidermal layer were found as shown in Fig. 5.8. Although under all the tested parameters some damage to the SC layer was consistent and repeatable, epidermal breach was not consistent when the nominal velocity was <3 ms^{-1} or with drop masses 7 and 17 g.

A correlation between the MB and histology images is shown in Fig. 5.9. By comparing the MB stained images in Fig. 5.9a, d and the histological cross-sections in Fig. 5.9b–c, e–f a more holistic insight in to the interaction between MNs and skin is obtained; although the MB stains confirmed successful penetration of the SC with a large number of MNs at 2 ms^{-1} with the 36 g drop mass (Fig. 5.9d), the histology images (Fig. 5.9e, f) show that there is only minimal disruption

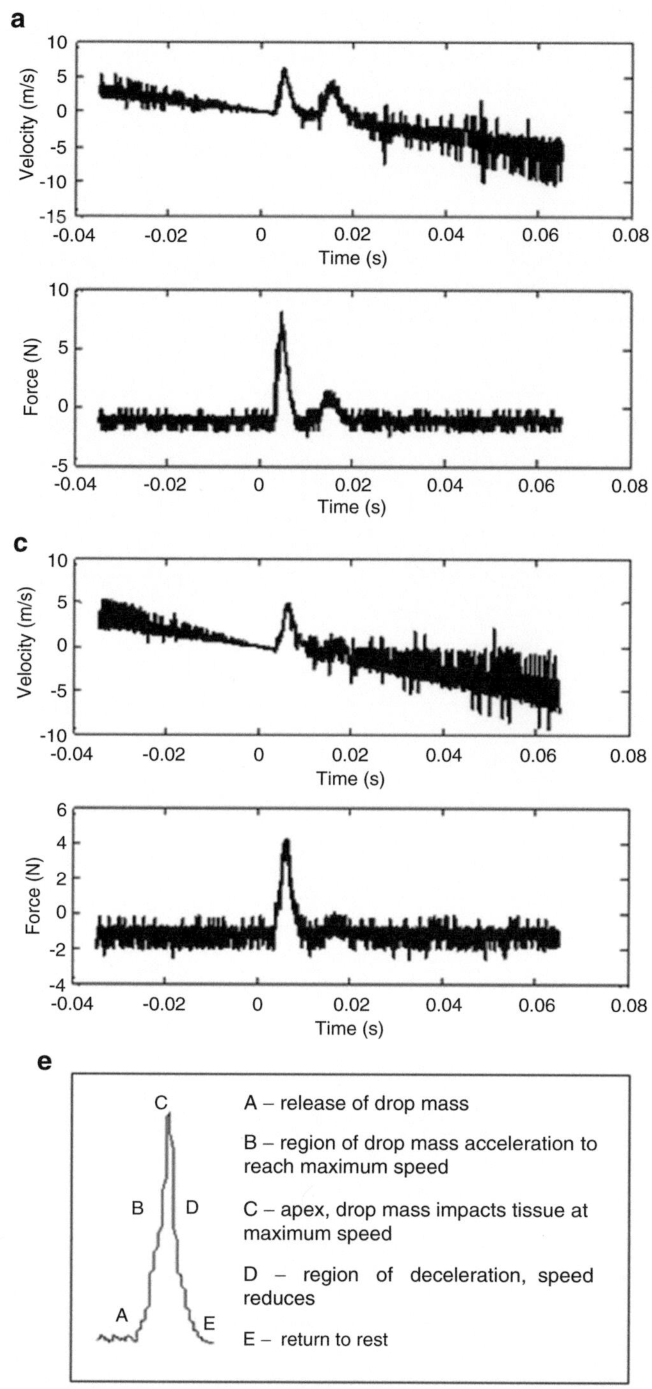

Fig. 5.6 Velocity–time plots and force–time plots for impact tests of MNs on the skin model at constant drop mass 36 g and various nominal velocities. (**a**) 5 ms^{-1}, (**b**) 4 ms^{-1}, (**c**) 3 ms^{-1}, (**d**) 2 ms^{-1} and (**e**) a description of points along the curve

to the SC, In contrast, there is a more substantial breach to the epidermal layer at 3 ms^{-1} (Fig. 5.9b, c). Additional tests validated this result i.e. that a 3 ms^{-1} nominal velocity with the 36 g drop mass (equating to a force of 3.35–4.17 N as shown in Fig. 5.6c) were found to be the ideal parameters for achieving reproducible and predictable MN penetration of skin into the epidermal layer.

The setup used in this study has allowed both force and velocity to be varied. This is in contrast to other studies which have tended to use a fixed force [23] or only considered the application velocity [22]. In terms of penetration efficiency, others studies in the literature have largely relied on MB or Trypan blue staining to assess penetration efficiency [22, 23],

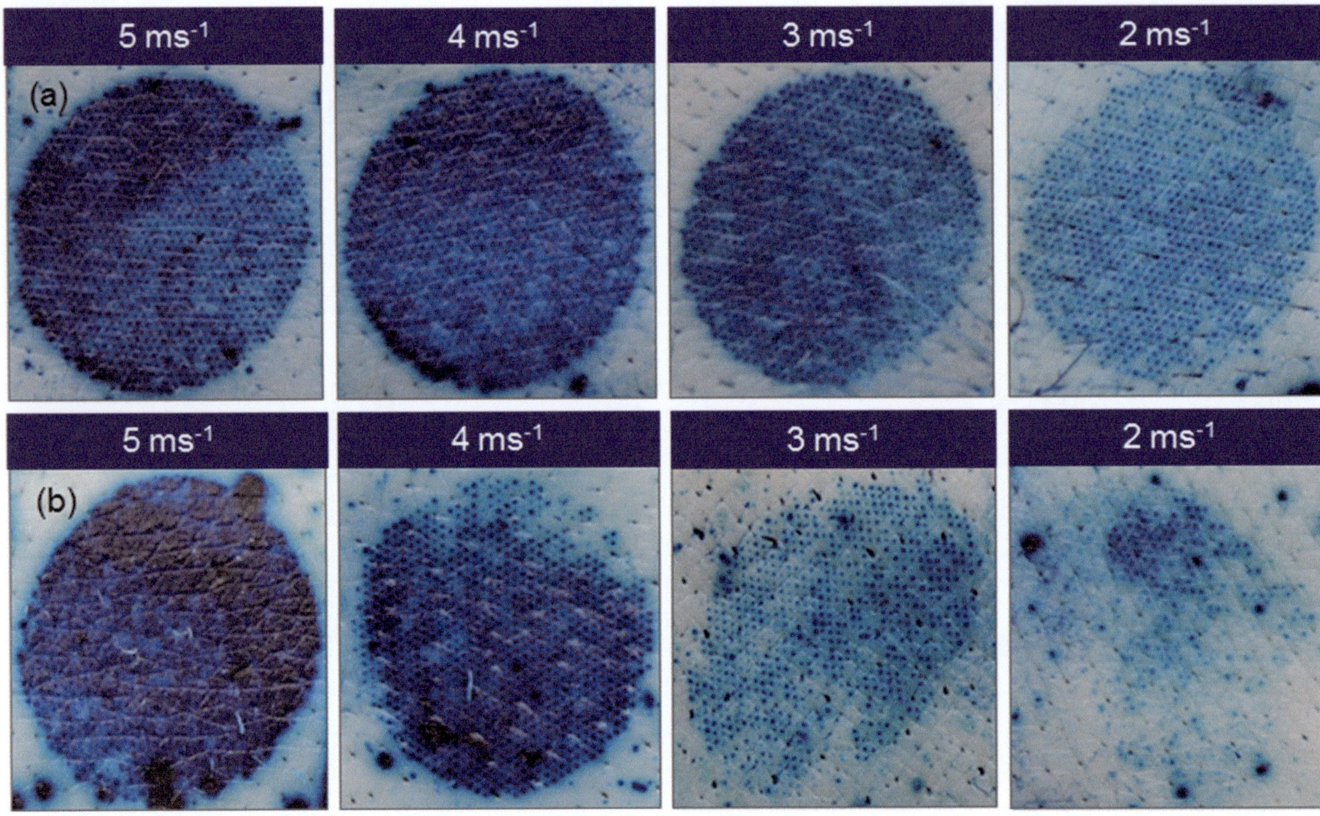

Fig. 5.7 Example images of MB stained skin following impact tests. (**a**) Representative images obtained at the various nominal velocities with the 36 g drop mass. (**b**) Representative images obtained at the various nominal velocities with the 17 g drop mass

however as shown in Fig. 5.9, a combination of such staining with histology is required to fully assess the extent of penetration. The additional novelty of this approach outlined in this paper is the use of a skin model which incorporates the subcutaneous tissue layer and muscle underneath the skin.

5.4 Conclusions

In this study, a dropped weight impact test rig was found to be suitable for determining optimal velocities and forces for reproducible and predictable insertion of high-density MNs into skin in vitro. The optimal velocity and force parameters for MN treated skin ranged from 4.52 to 4.8 ms^{-1} (nominal velocity of 3 ms^{-1}) and 3.35–4.17 N. Histology imaging demonstrated that in some cases although the SC layer of skin was disrupted, the MNs did not fully penetrate the epidermis.

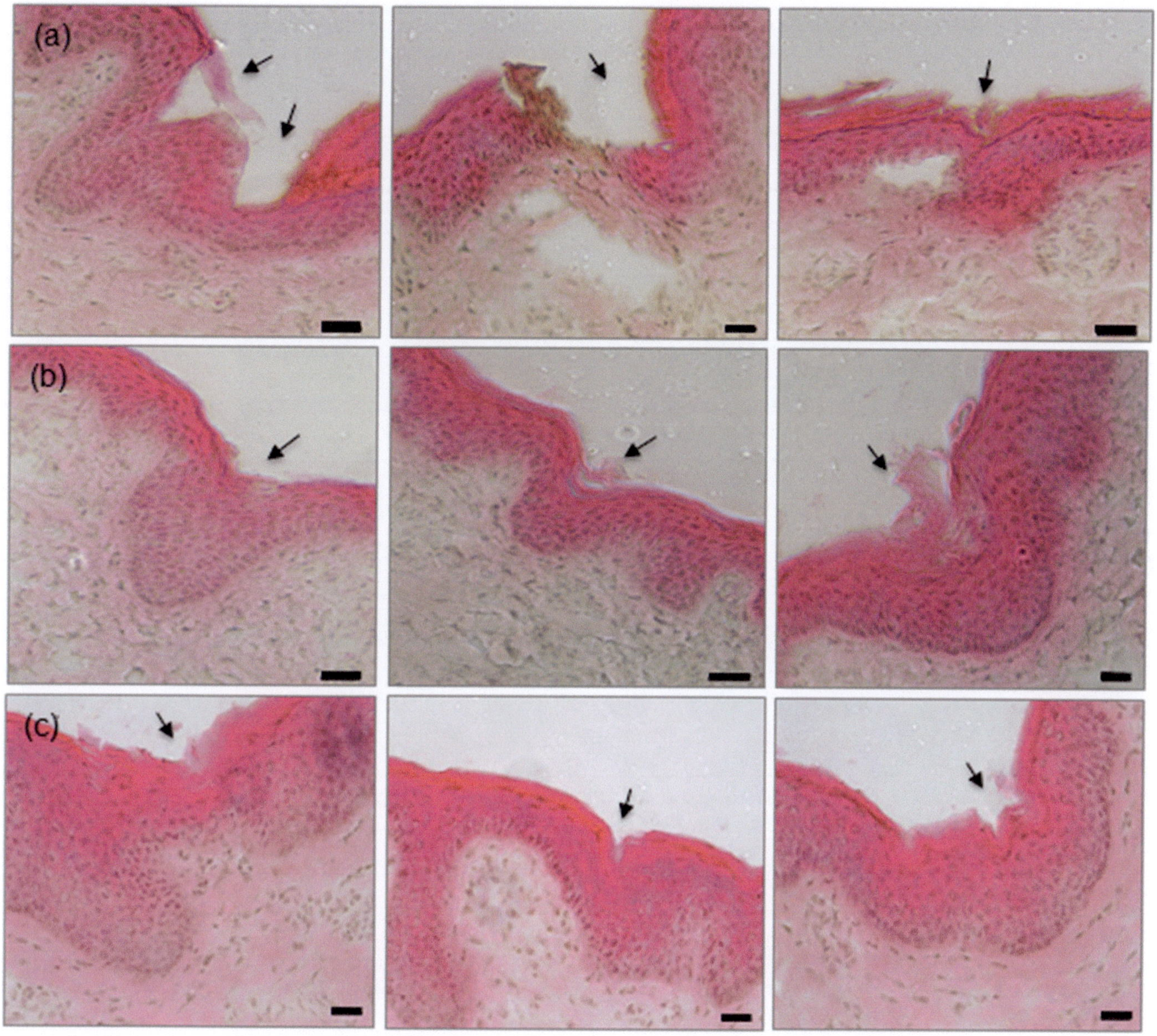

Fig. 5.8 Histology images of skin following MN application at a constant nominal velocity (2 ms^{-1}) under various loads. (**a**) 36 g, (**b**) 17 g and (**c**) 7 g. Scale bar represents 100 μm. *Arrows* highlight the location where the MNs have either disrupted or penetrated the SC

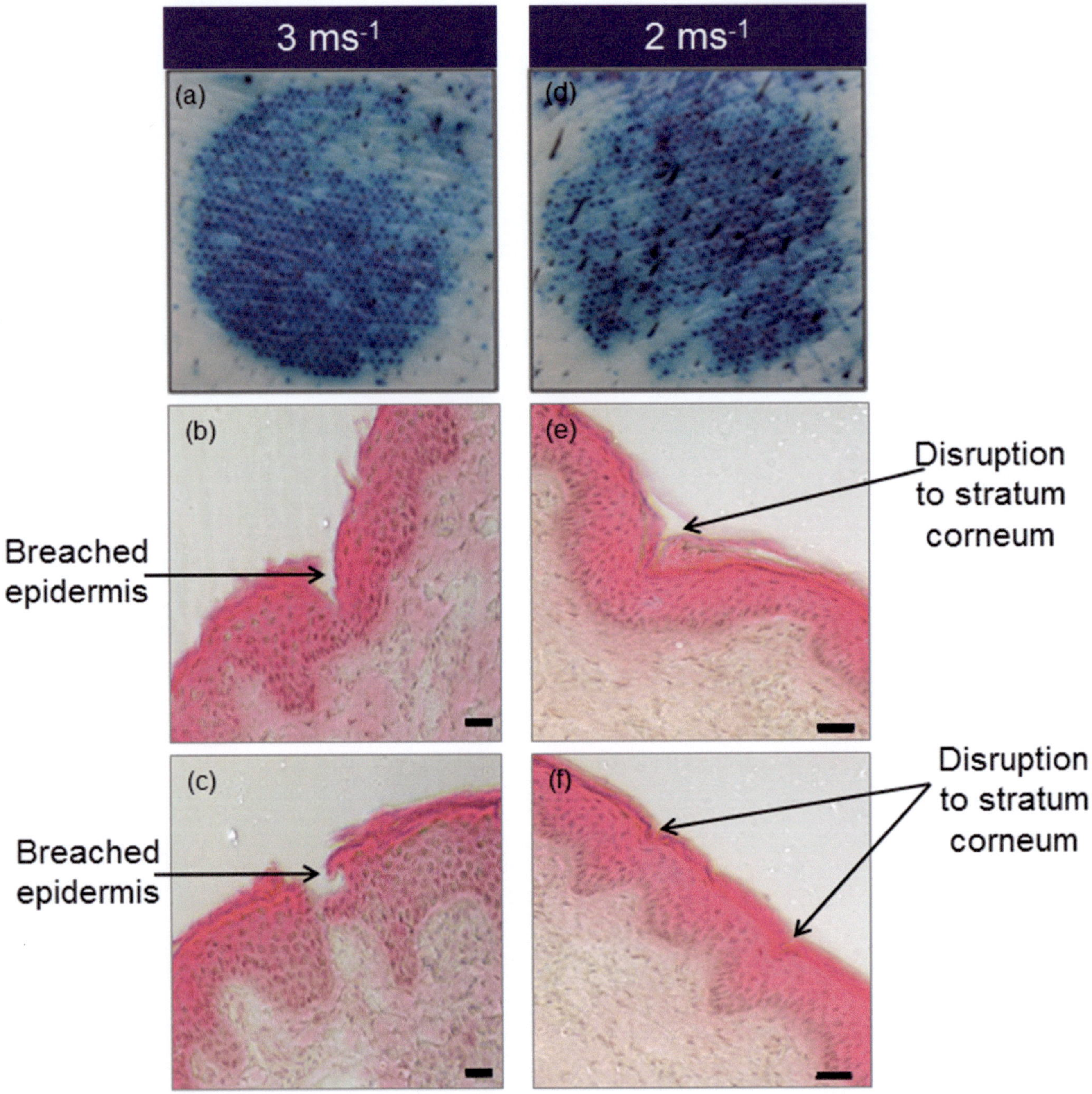

Fig. 5.9 The influence of velocity on perforation quality (MB staining) and perforation depth (H&E staining), with the 36 g drop mass at a velocity of 3 ms^{-1}. (**a**) MB image. (**b**) and (**c**) Example histology images, and a velocity of 2 ms^{-1}. (**d**) MB image. (**e**) and (**f**) Example histology images. Scale bar represents 100 μm

Acknowledgements This work was funded by Centre for Global Eco-Innovation PhD studentship in conjunction with Renephra Ltd. The authors would like to Derek Neary and Dr Rob Birch for assistance with the experimental setup. RA is grateful to the Academy of Medical Sciences for providing travel funding to present this paper at the SEM Annual Conference 2015.

References

1. Pailler-Mattei, C., Bec, S., Zahouani, H.: In vivo measurements of the elastic mechanical properties of human skin by indentation tests. Med. Eng. Phys. **30**(5), 599–606 (2008)
2. Pegoraro, C., MacNeil, S., Battaglia, G.: Transdermal drug delivery: from micro to nano. Nanoscale **4**(6), 1881 (2012)
3. Bremseth, D.L., Pass, F.: Delivery of insulin by jet injection: recent observations. Diabetes Technol. Ther. **3**, 225–232 (2001)
4. Karpozilos, A., Pavlidis, N.: The treatment of cancer in Greek antiquity. Eur. J. Cancer **40**(14), 2033–2040 (2004)
5. Riviere, J.E., Heit, M.C.: Electrically-assisted transdermal drug delivery. Pharm. Res. **14**, 687–697 (1997)
6. Haedersdal, M., et al.: Fractional CO(2) laser-assisted drug delivery. Lasers Surg. Med. **42**(2), 113–122 (2010)
7. Bal, S.M., et al.: In vivo assessment of safety of microneedle arrays in human skin. Eur. J. Pharm. Sci. **35**(3), 193–202 (2008)
8. Mukerjee, E.V., et al.: Microneedle array with integrated microchannels for transdermal sample extraction and in situ analysis. In: 12th International Conference on Solid State Sensors, Actuators and Mircosystems, Boston (2003)
9. Henry, S., et al.: Microfabricated microneedles a novel approach to transdermal drug delivery. J. Pharm. Sci. **87**, 922–925 (1998)
10. Zahn, J.D., et al.: Microfabricated polysilicon microneedles for minimally invasive biomedical devices. Biomed. Microdevices **2**, 295–303 (2000)
11. Park, J.-H., Allen, M.G., Prausnitz, M.R.: Biodegradable polymer microneedles: fabrication, mechanics and transdermal drug delivery. J. Control. Release **104**(1), 51–66 (2005)
12. Martanto, W., et al.: Transdermal delivery of insulin using microneedles in vivo. Pharm. Res. **21**, 947–952 (2004)
13. Lee, J.W., et al.: Dissolving microneedle patch for transdermal delivery of human growth hormone. Small **7**(4), 531–539 (2011)
14. Arora, A., Prausnitz, M.R., Mitragotri, S.: Micro-scale devices for transdermal drug delivery. Int. J. Pharm. **364**(2), 227–236 (2008)
15. Prausnitz, M.R.: Overcoming the skin's barrier: the search for effective and user-friendly drug delivery. Diabetes Technol. Ther. **3**, 233–236 (2001)
16. Olatunji, O., Das, D.B., Al-Qallaf, B.: Simulation based optimisation of microneedle geometry to improve drug permeability in skin, In: Proceedings of the 7th Industrial Simulation Conference' 2009 (ISC'2009), 293–300, Loughborough (2009)
17. van Gerwen, D.J., Dankelman, J., van den Dobbelsteen, J.J.: Needle-tissue interaction forces—a survey of experimental data. Med. Eng. Phys. **34**(6), 665–680 (2012)
18. Ni Annaidh, A.N., et al.: A combined experimental and numerical study of stab-penetration forces. Forensic Sci. Int. **233**(1–3), 7–13 (2013)
19. Shergold, O.A., Fleck, N.A.: Experimental investigation into the deep penetration of soft solids by sharp and blunt punches, with application to the piercing of skin. J. Biomech. Eng. **127**(5), 838 (2005)
20. Sivamani, R.K., Liepmann, D., Maibach, H.I.: Microneedles and transdermal applications. Expert Opin. Drug Deliv. **4**, 19–25 (2007)
21. Donnelly, R.F., et al.: Optical coherence tomography is a valuable tool in the study of the effects of microneedle geometry on skin penetration characteristics and in-skin dissolution. J. Control. Release **147**(3), 333–341 (2010)
22. Verbaan, F.J., et al.: Improved piercing of microneedle arrays in dermatomed human skin by an impact insertion method. J. Control. Release **128**(1), 80–88 (2008)
23. van der Maaden, K., et al.: Impact-insertion applicator improves reliability of skin penetration by solid microneedle arrays. AAPS J. **16**(4), 681–684 (2014)
24. Shergold, O.A., Fleck, N.A., Radford, D.: The uniaxial stress versus strain response of pig skin and silicone rubber at low and high strain rates. Int. J. Impact Eng. **32**(9), 1384–1402 (2006)
25. Ryckman, R.A., Powell D.A., Lew A.: Ballistic penetration of Perma-Gel, In: Proceedings of the 17th American Physical Society Topical Conference on Shocked Compression of Condensed Matter, 143–148, American Institute of Physics, Chicago (2011)
26. Li, G., et al.: Microchannels created by sugar and metal microneedles: characterization by microscopy, macromolecular flux and other techniques. J. Pharm. Sci. **99**(4), 1931–1941 (2010)

Chapter 6
DIC Measurements of the Human Heart During Cardiopulmonary Bypass Surgery

Mikko Hokka, Nikolas Mirow, Horst Nagel, Sebastian Vogt, and Veli-Tapani Kuokkala

Abstract Image-based measurements of the deformation of the human heart can be very useful to the surgeon, when assessing the condition and functioning of the patient's heart. Digital image correlation can provide fast and accurate information about the deformation and motion of the surface of the heart. The deformation measurements can be visualized with colors allowing easy interpretation of the results, which makes this technique even more suitable for use in the operating room. Digital image correlation, however, requires either a natural or an artificial surface pattern with high contrast. The surface of the heart is wet, smooth, and has only a minimal contrast pattern, which cannot easily be improved with artificial markers. This preliminary feasibility study, however, shows that despite the practical and theoretical problems, DIC can provide useful data on the deformation of the human heart during cardiopulmonary bypass surgery. The results show that the natural patterns of the right atrium and ventricle are sufficient for DIC analysis, but significantly better results could be obtained with higher contrast artificial patterns.

Keywords Digital Image Correlation • In-vivo measurements • Human heart • Natural pattern

6.1 Introduction

Measurements of the wall-motion of the right ventricle (RV) can be very useful in assessing the patient's condition during open heart surgery. The deformation or motion of the heart muscle's (myocardium) surface can be used for estimating the functioning of the heart, especially during the weaning period after surgical repair operations. During the operation the heart is typically stopped from beating and the extracorporeal circulation (ECC) provides the blood transport. However, despite being protected by cardioplegic drugs the heart suffers from ischemic arrest during the surgical repairs and the muscle can be damaged. When the blood flow is returned to the coronary arteries, the heart slowly recovers its normal function. During this period correct regulation of intravasal blood volume is important. Too little blood does not provide sufficient perfusion for the organs and leads to a risk of ischemia. Too much volume, on the other hand, easily leads to too much pressure for the still weak heart causing unwanted expansion and eventual failure of the myocardium. Therefore, monitoring the volume load, deformation, and contraction patterns of the heart are very important during the weaning period. Previously, echocardiography has been used for the measurements of the wall-motion [1]. During the operation, an ultrasonic sensor is inserted into the esophagus of the patient. However, due to technical reasons and reasons arising from the human anatomy, echocardiography cannot very easily be applied to the measurements of the right ventricle or the atria. The sensor is mainly used for imaging of the left ventricle, which is not fully visible to the surgeon. The clinical difficulties are discussed in more details for example by Teske et al. [2]. For these reasons, new methods and techniques are being developed for the measurements of the RV deformation behavior. For example, tissue Doppler techniques [3], magnetic resonance tagging [4] and finite element simulations [5] can be used to calculate the axial and longitudinal deformations. However, these approaches are currently too slow to provide on-line data during the weaning period and, therefore, cannot be used for the on-line assessment of the functioning of the patient's heart.

M. Hokka (✉) • V.-T. Kuokkala
Department of Materials Science, Tampere University of Technology, Tampere, Finland
e-mail: mikko.hokka@tut.fi

N. Mirow • S. Vogt
Heart Surgery, Universitätsklinikum Gießen und Marburg GmbH, Marburg, Germany

H. Nagel
LaVision LTD, Göttingen, Germany

S.A. Tekalur et al. (eds.), *Mechanics of Biological Systems and Materials, Volume 6*,
Conference Proceedings of the Society for Experimental Mechanics Series, DOI 10.1007/978-3-319-21455-9_6

DIC has been successfully used for the deformation measurements of artificial soft tissues by Moerman et al. [6], and Han et al. [7] used DIC for images obtained with ultrasound to analyze the breast cancer tissue. Libertiaux et al. [8] studied the brain tissue and Gao [9] the liver tissues with DIC. Overall, the use of DIC for analyzing the mechanical behavior of very soft tissues is challenging, because in most cases normal patterning methods cannot be used. The paint might not stick if the surface is wet or greasy, and it also may significantly change the mechanical properties, such as stiffness, of the surface of the soft tissue. Therefore, various tissue dyes, particles, and gels have been used for patterning. However, no work has been reported concerning in-vivo deformation measurements of the human heart during open heart surgery. In this work, a commercial digital image correlation system (StrainMaster) from LaVision Inc. was used for a preliminary feasibility study. This paper reports the experimental procedure for obtaining the images and the preliminary deformation analysis of the human heart during a cardiopulmonary bypass surgery.

6.2 Experimental

The imaging took place at the Universitätsklinikum Marburg (Germany). Prior to the operation, the measurement equipment was set up in the operating room in only about 45 min. The operating room was continuously used for operations, and it was not available for preparations. Due to this tight schedule the positioning and focusing of the cameras were not optimal. Unfortunately this later affected the calibration quality and accuracy of the displacement and strain calculations. In the operating room the cameras were fixed to a horizontal rail above and slightly in front of the patient's chest so that a good unobstructed optical path from the cameras to the heart was established. Two 5Mpix E-lite cameras with 50 mm Nikon objective lenses were used to obtain images of the beating heart at a rate of ten images per second. No artificial pattern was applied onto the surface of the heart.

Prior to the operation, images of the LaVision two-level calibration target were obtained and a preliminary pinhole model calibration was formed. However, the exact distance from the calibration plate to the heart was not known at the time when the calibration images were taken. Therefore, the distance from the operating room floor was used as a reference measure. The distance from the calibration target to the operating room floor was measured with a tape measure when the images were taken. The distance from the heart to the operating room floor was more difficult to measure due to various practical reasons. It was therefore simply estimated using a similar tape measure but without touching the patient. This of course was not very accurate, but using this distance as a start point for the Z=0 mm plane, the calibration could be iteratively improved. The first image pair was first converted to a 3D image using the original pinhole calibration, and the calibration (Z=0 mm plane) was then transformed to the surface of the heart obtained from the first image pair. Normally this is done only once if small errors in the Z-level need to be corrected, but in this case the transformation was iterated 3–4 times to obtain a reasonable calibration with a standard deviation of 0.458 pixels, where one pixel corresponds to a distance of 146.8 μm (Fig. 6.1).

The images were analyzed using DaVis (v.8.20) software from LaVision. The displacements on the surface of the heart were obtained by tracking the movement of square image subsets of 121×121 pixels using a step size of one pixel. With the very poor contrast pattern of the surface of the heart, the accuracy of the strain calculations increases with a large subset that contains more data for the image correlation. Also, much larger areas can be analyzed with a larger subset compared to the analysis based on smaller subsets. The step size of one pixel gives the displacement for each pixel but does not significantly increase the spatial displacement resolution compared to to the use of a larger step size and smoothing between the steps. However, the area growth algorithm used in the DaVis software is significantly more stable with small step sizes, and therefore the calculations require less seed points. The use of a large subset and a small step size, however, drastically increase the calculation time, which in practice limits the analysis to a few heart beats only.

6.3 Results

Figure 6.2 shows a series of analyzed images with a time step of 0.1 s. The images show the right side of the heart, whereas the left side of the heart cannot be seen. The overlaid colors represent the displacement of the surface of the heart in the Z-direction relative to the reference (first) image. The images clearly show that both the right ventricle and the atrium first expand in the Z-direction but start to contract after 0.3 s (Fig. 6.2d). The deformation is not uniform after this point, as the center of the analyzed area clearly moves down (blue color), while the areas on the edges show positive Z-displacement. The heart muscle recovers towards its original form in Fig. 6.2e, after which the next beat starts with a strong positive Z-displacement. The total surface displacement from the most contracted configuration to the most expanded configuration is close to six millimeters. This maximum displacement occurs repeatedly at the center of the analyzed region.

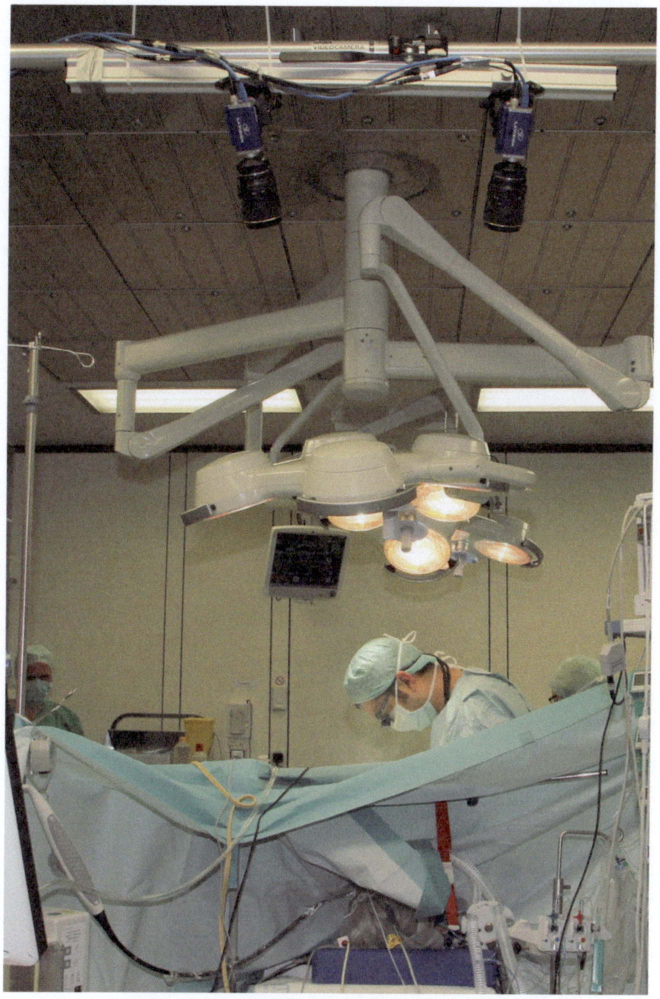

Fig. 6.1 Positioning of the cameras in the operating room

Figure 6.3 shows the same series of images as Fig. 6.2 but now for the horizontal (X-direction) displacement. The displacements in the Y-direction were also studied, but they were rather constant in the analyzed region as a function of time. However, the X-direction displacement shows clear regional differences. In Fig. 6.3, the left side of the analyzed region moves in the positive direction (to the right), whereas the right side of this region shows negative displacements (moves left). Maximum displacements are found in Fig. 6.3f.

Figure 6.4 shows two virtual extensometers placed on the images of the heart. The extensometers are placed approximately along the circumferential direction (Fig. 6.4a) and in the longitudinal direction (Fig. 6.4b). The engineering strains along the strain gage lines are shown in Fig. 6.4c for several heart beats. For the first two heart beats the strains in both directions are quite similar. The maximum compressions relative to the reference image are between 4 and 4.5 %, while the tensile strains are less than 1 %. However, the third and fourth beats are clearly different from the first two, and the deformations in the circumferential and longitudinal directions differ quite much from each other: in the longitudinal direction the heart expands more, while in the circumferential direction the heart contracts more strongly.

6.4 Discussion

The heart muscle has a fairly uniform color and its surface provides a very weak contrast. Compared to typical Digital Image Correlation applications, the overall sterile environment and the requirements for a sterile, biocompatible and preferably removable pattern are very challenging. In addition to this, the short preparation time prevented the use of optimal optics. The images obtained with 50 mm lenses contain quite much redundant information and only a small fraction of the 5 M

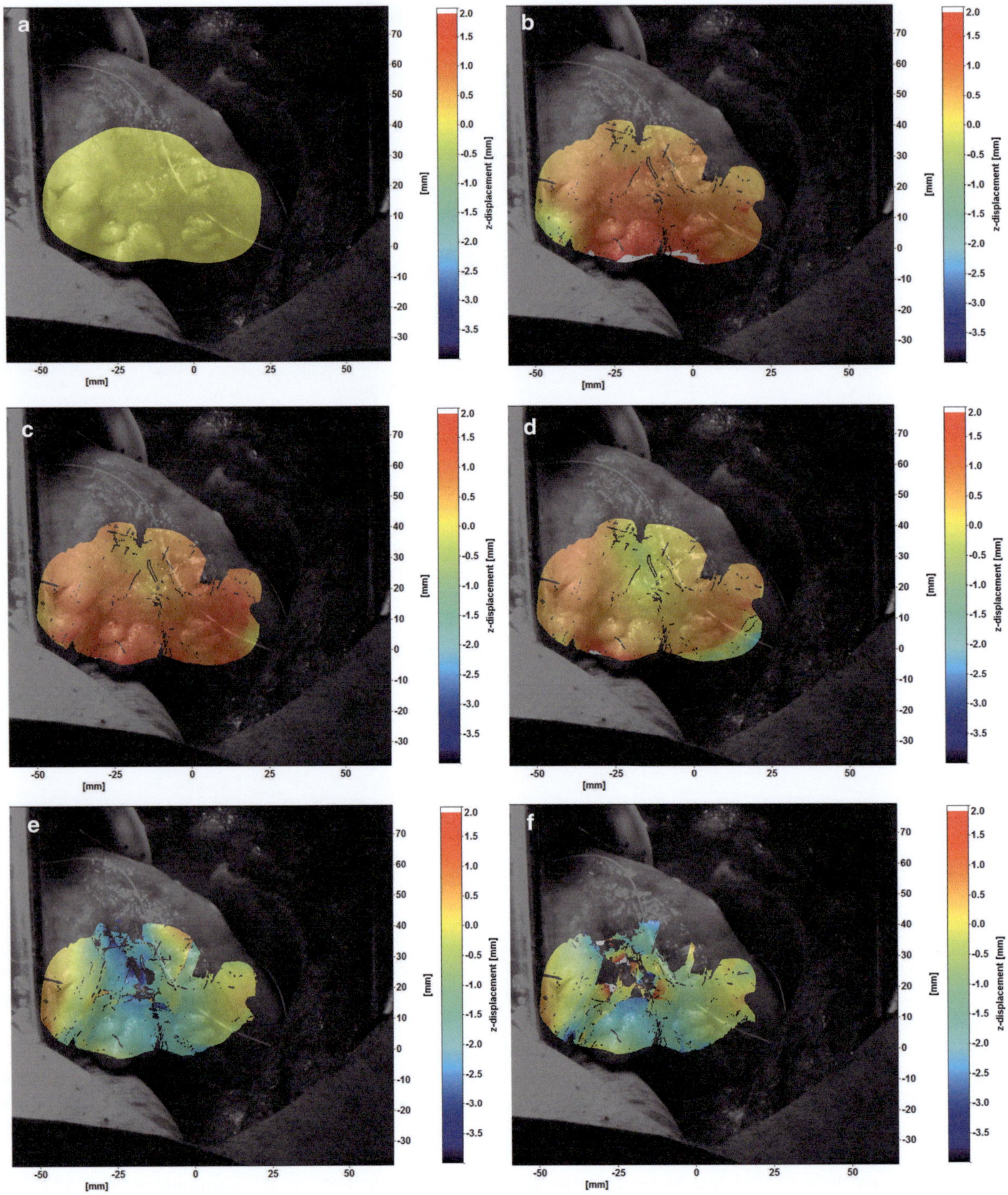

Fig. 6.2 Displacement of the surface of the heart in the Z-direction in 0.1 s steps

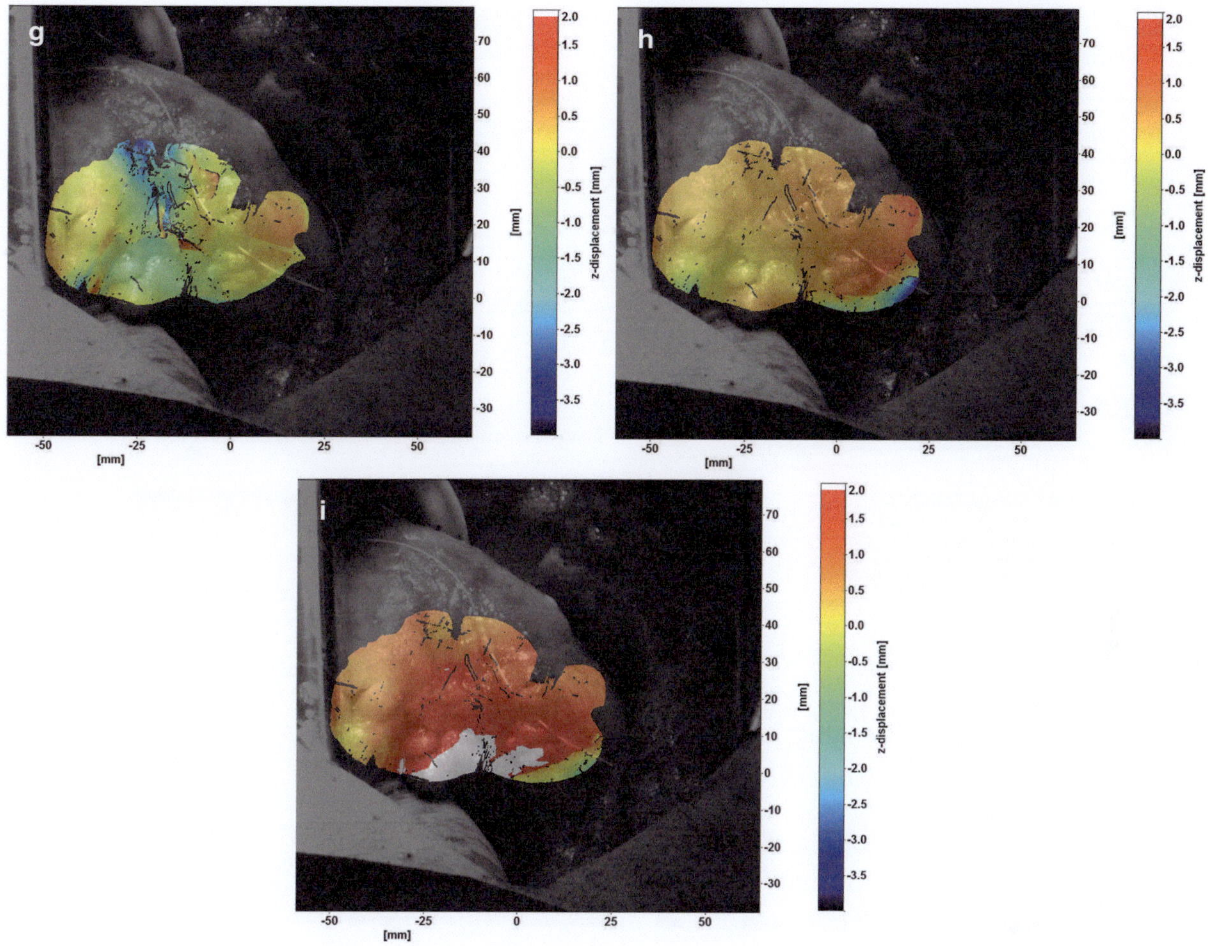

Fig. 6.2 (continued)

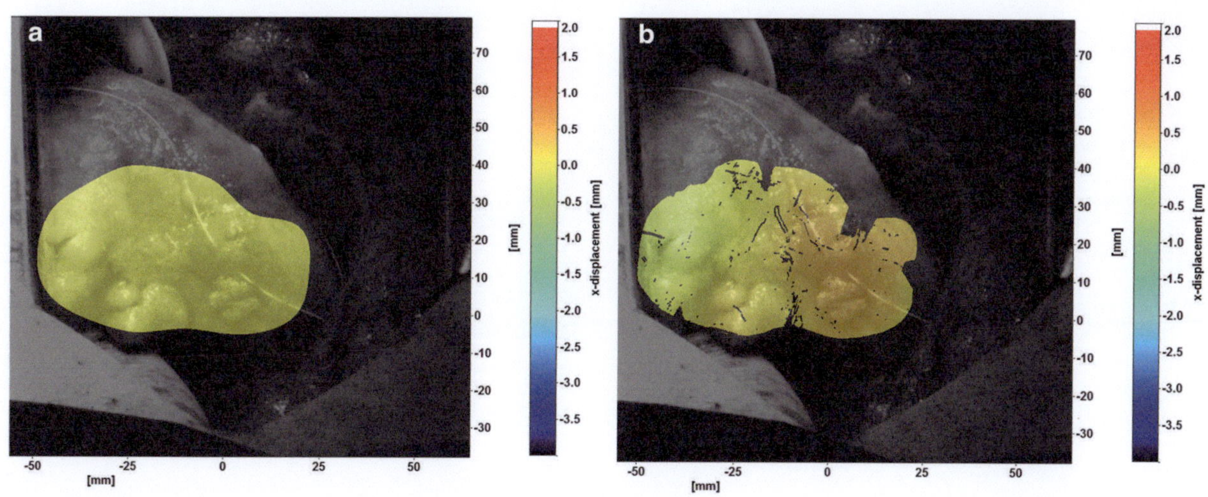

Fig. 6.3 Displacement of the surface of the heart in the X-direction in 0.1 s steps

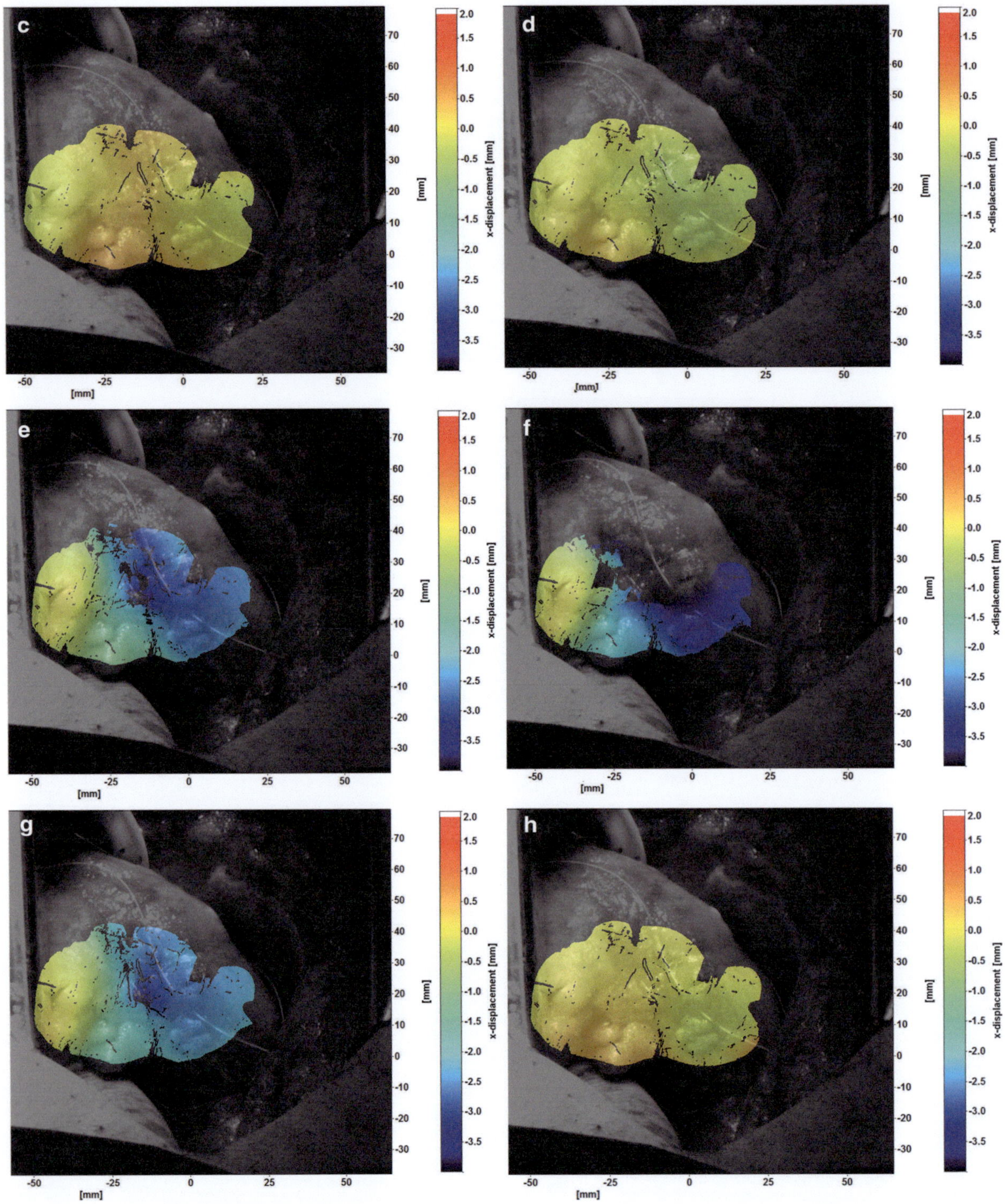

Fig. 6.3 (continued)

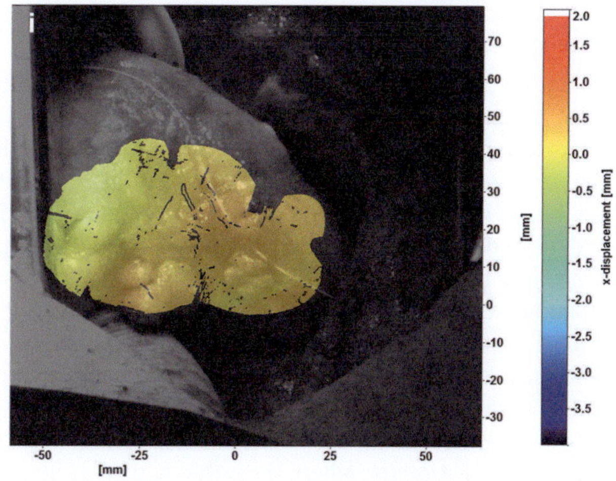

Fig. 6.3 (continued)

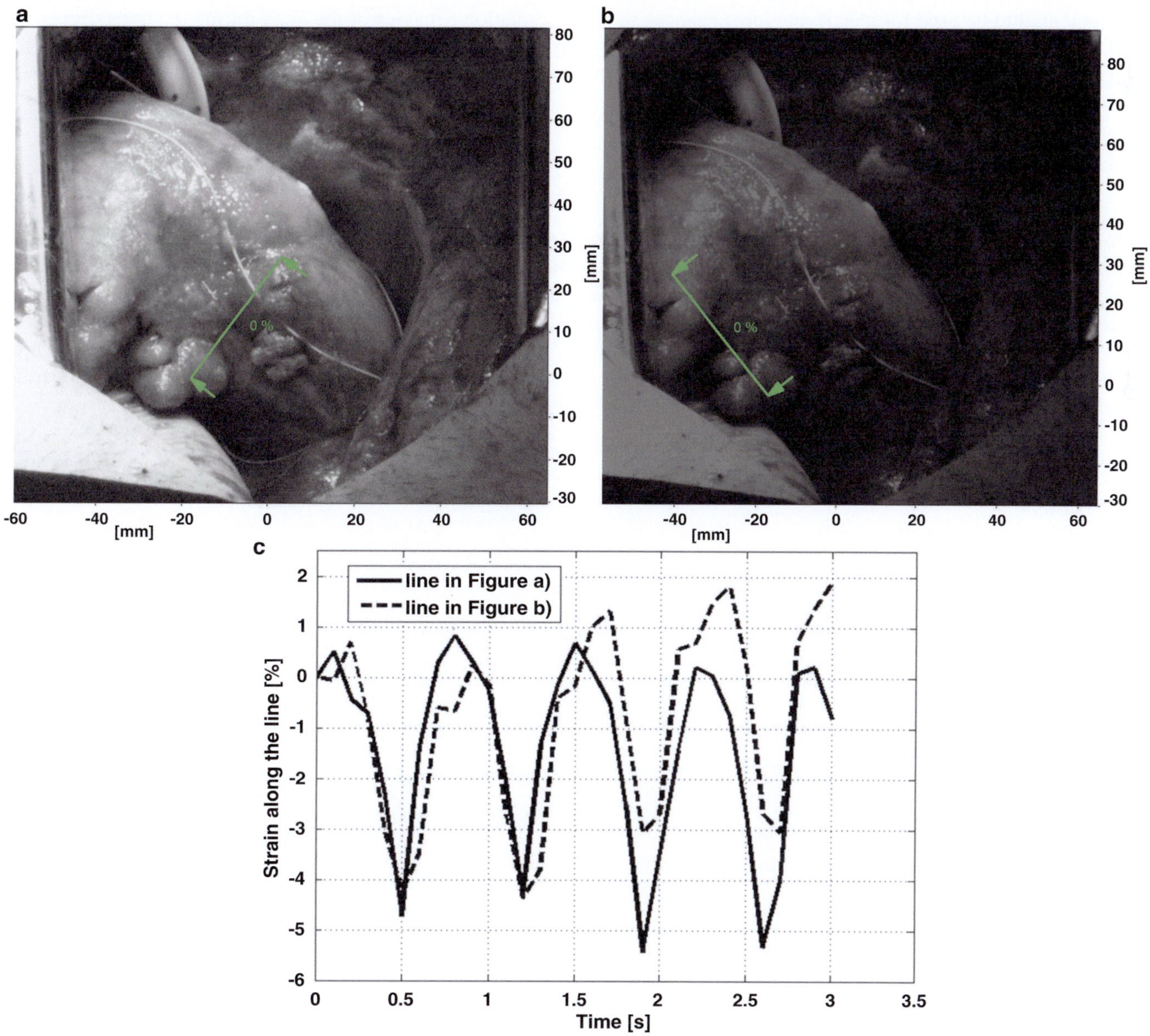

Fig. 6.4 Virtual extensometers used to obtain the (**a**) circumferential and (**b**) longitudinal engineering strains. (**c**) shows the strains as a function of time

pixels cover the heart. The spatial resolution of the DIC analysis could be significantly improved with optimized optics and especially by the use of a good contrast pattern on the surface of the heart. For young and athletic patients, the heart muscle is reddish and may have some natural texture from the muscle tissue. However, for older patients the heart muscle is covered by a yellowish layer of fat, which has an essentially uniform color and very low natural contrast. Several scientists have analyzed the deformation of soft tissues with improved artificial patterns. For example, Miri et al. [10] used commercial tissue dyes, Myers et al. [11] used powders, and Thompson et al. [12] used gels mixed with powders to enhance the contrast. Lionello et al. [13] developed tools specially designed for enhancing the contrast patterns of soft tissues. However, none of these methods can be applied to human heart surgery since the pattern must be fully nontoxic, biocompatible, and sterile. However, natural patterns have been used to analyze soft tissues also before. Gao et al. [9] used 2D digital image correlation to study the deformation of liver tissue using only the natural pattern of the tissue. The liver tissue has strong texture which in that case gave good correlation results. In the current study, however, the texture on the surface of the heart is not as strong and the contrast in the images remains poor. Despite this, the stereo reconstruction errors remain at a reasonable level (less than 0.5 pixels) except for the areas that deform strongly, where the errors can be as high as 4 pixels.

The deformation of the heart is very complex. Some regions expand while the others contract, resulting in very complex and rapidly changing deformation gradients. However, based on the analyzed images, DIC can resolve displacements and deformation of the surface of the human heart. The method can be used to compare the deformations during the analyzed heartbeats and between the different regions of the heart even with a very poor image contrast pattern. Therefore, DIC has clear potential to be used for identifying local changes in the deformation behavior of the heart during the heart operation. This can be very useful for the surgeon and the anesthesiologists when comparing the functioning of the heart in the beginning of the operation and in the later stages, especially during the weaning from ECC.

6.5 Summary

Deformation of the right side of the human heart was studied using digital image correlation during a cardiopulmonary bypass surgery. Images of the heart were analyzed based on the natural surface pattern of the heart. The displacements in the Z- and X- directions were found to be clearly different in the different analyzed areas, whereas the displacements in the Y-direction were significantly more homogeneous. Also engineering strains in the longitudinal and circumferential directions were obtained using virtual extensometers. The deformations were observed to vary for different heartbeats. The preliminary results of this work show that DIC can be used for obtaining detailed information of the functioning of the heart at high spatial resolution.

References

1. La Gerche, A., Ruxandra, J., Voigt, J.: Right ventricular function by strain echocardiography. Curr. Opin. Cardiol. **25**, 430–436 (2010)
2. Teske, A., Cox, M., De Boeck, B.: Echocardiographic tissue deformation imaging quantifies abnormal regional right ventricular function in arrythmogenic right ventricular dysplasia/cardiomyopathy. J. Am. Soc. Echocardiogr. **22**, 920–927 (2009)
3. Pauliks, L., Valdes-Cruz, L., Perryman, R., Scholl, F.: Right ventricular wall-motion changes after infant open heart surgery—a tissue Doppler study. Echocardiography **31**, 209–217 (2014)
4. Ibrahim, E.: Myocardial tagging by cardiovascular magnetic resonance: evolution of techniques—pulse sequences, analysis algorithms, and applications. J. Cardiovasc. Magn. Reson. **13**, 36 (2011)
5. Dorri, F., Niederer, P.F., Lunkenheimer, P.P.: A finite element model of the human left ventricular systole. Comput. Methods Biomech. Biomed. Engin. **9**, 319–341 (2006)
6. Moerman, K., Holt, C., Evans, S., Simms, C.: Digital image correlation and finite element modelling as a method to determine mechanical properties of human soft tissue in vivo. J. Biomech. **42**, 1150–1153 (2009)
7. Han, Y., Kim, D.-W., Kwon, H.: Application of digital image cross-correlation and smoothing function to the diagnosis of breast cancer. J. Mech. Behav. Biomed. Mater. **14**, 7–18 (2012)
8. Libertiaux, V., Pascon, F., Cescotto, S.: Experimental verification of brain tissue incompressibility using digital image correlation. J. Mech. Behav. Biomed. Mater. **4**, 1177–1185 (2011)
9. Gao, Z., Desai, J.: Estimating zero-strain states of very soft tissue under gravity loading using digital image correlation. Med. Image Anal. **14**, 126–137 (2009)
10. Miri, A., Barthelat, F., Mongeau, L.: Effects of dehydration on the viscoelastic properties of vocal folds in large deformations. J. Voice **26**, 688–697 (2012)

11. Myers, K., Coudrillier, B., Boyce, B., Nguyen, T.: The inflation response of the posterior bovine sclera. Acta Biomater. **6**, 4327–4335 (2010)
12. Thompson, M., Schell, H., Lienau, J., Duda, G.: Digital Image correlation: a technique for determining local mechanical conditions within early bone callus. Med. Eng. Phys. **29**, 820–823 (2007)
13. Lionello, G., Sirieix, C., Baleani, M.: An effective procedure to create speckle pattern on biological soft tissue for digital image correlation measurements. J. Mech. Behav. Biomed. Mater. **39**, 1–8 (2014)

Chapter 7
Comparing 2D and 3D Digital Image Correlation for an Inflation Test

Barbara J. Murienne and Thao D. Nguyen

Abstract 3D digital image correlation (DIC) is widely used in biomechanics to measure the deformation of 3D structures. However, 3D-DIC implementation can be challenging under specific conditions, such as high magnification (i.e. low depth of focus) or environment with limited access for two-angled cameras. Here, we aimed to compare 2D-DIC and 3D-DIC for the same inflation test, to evaluate 2D-DIC as an alternative to 3D-DIC under those conditions. A two-camera stereovision system acquired top-down images of a membrane inflated vertically to 5.41 kPa (reference pressure), then to 7.87 kPa (deformed pressure), while a single camera system simultaneously recorded side images. 2D-DIC and 3D-DIC were used to calculate vertical (out of membrane plane) and horizontal (in membrane plane) displacements, and meridional strain. Under inflation, the variability in displacements and strains was larger than under static conditions for both DIC methods. 2D-DIC showed a smaller variability than 3D-DIC, especially in the vertical displacements, and the absolute difference between the averaged 3D-DIC and 2D-DIC data was in the range of the 3D-DIC variability. The meridional strain uncertainty was similar between both DIC methods. Those findings suggests that 2D-DIC might be an alternative to 3D-DIC for inflation under certain conditions.

Keywords Digital image correlation • 3D full-field measurements • 2D profile measurements • Inflation • Mechanical testing

7.1 Introduction

DIC is a non-contact method used to characterize the local mechanical behavior of specimens under deformation. In recent years, this technique has been increasingly used to study biological materials under physiological conditions [1–10]. Due to the 3D nature of the shape and deformation of biological materials in vivo, 3D-DIC is commonly used to investigate their mechanical behavior *ex-vivo*. In fact, 2D-DIC is not recommended for 3D specimens as well as planar specimens undergoing large amounts of out-of-plane translation and rotation [11].

In 2D-DIC, a single camera acquires images of the surface of a planar specimen, showing a random intensity pattern, during deformation. Each reference subset within the reference image (typically the first image) is then compared to deformed subsets within a deformed image (any of the subsequent images) to find the target subset. The target subset is the subset in the deformed image that shows the maximum pattern similarity with the reference subset, based on a maximum cross-correlation criterion or minimum sum-squared difference criterion [12]. The correlation accuracy mainly depends on the camera noise, whether the camera and specimen surface are parallel, image distortion, illumination conditions, speckle pattern and matching process (subset size, shape function, interpolation scheme, correlation criterion) [12]. 2D displacements vectors for each point on the surface of the specimen are finally obtained. This process is repeated for each deformed image to obtain the 2D displacements of the specimen surface throughout the deformation.

In 3D-DIC, two cameras acquire images of the surface of the specimen simultaneously. The 3D algorithm first correlates the images of the reference image pair and reconstructs the specimen 3D geometry using the camera calibration parameters and triangulation. Calibration parameters contain information about the camera position in space and with respect to each

B.J. Murienne, M.Sc. (✉) • T.D. Nguyen, Ph.D.
Department of Mechanical Engineering, Johns Hopkins University,
Latrobe Hall 200, 3400 N. Charles St., Baltimore, MD 21218, USA
e-mail: bmurien1@jhu.edu

S.A. Tekalur et al. (eds.), *Mechanics of Biological Systems and Materials, Volume 6*,
Conference Proceedings of the Society for Experimental Mechanics Series, DOI 10.1007/978-3-319-21455-9_7

other. The algorithm then tracks the displacements of points on the surface of the specimen by correlating the images of the reference pair with the images of any subsequent pair [13]. 3D displacements vectors are finally obtained for each point on the surface of the specimen and by repeating the same process for all deformed image pairs, the 3D displacements are obtained for all time steps. Compared to 2D-DIC, which main source of error is the correlation error, 3D-DIC has a correlation and a 3D reconstruction error [14, 15]. The 3D reconstruction error in 3D-DIC depends on the camera positioning (stereo-angle) and calibration, and the objective focal length [14, 15].

Inflation testing is widely used to characterize the mechanical properties of biological materials [4–7, 9, 10, 16] as well as non-biological tubular materials [17] and thin films [18–22]. The use of 2D-DIC is often limited due to the complexity of the material structure and deformation however, some experimental conditions discourage or disallow the use of 3D-DIC. Those conditions include high magnification, and therefore a small depth of focus, and low accessibility for two-angled cameras, that may be caused by the need of a controlled environment during mechanical testing. Here, we aimed to compare 3D-DIC and 2D-DIC experimentally, for the same inflation experiment, to evaluate whether 2D-DIC could be a valid alternative to 3D-DIC under those specific conditions.

7.2 Methods

7.2.1 Specimen Preparation, Testing and Imaging

A 0.2032 ± 0.0508 mm thick latex membrane (Abrasion-resistant natural latex rubber film, 85995K13, McMaster-Carr, Princeton, NJ) was mounted on an acrylic holder with a 20.5 mm circular opening, secured down on a inflation chamber (Fig. 7.1a) and inflated from baseline pressure 0.28 to 7.87 kPa at 0.13 kPa/s, similar to the procedure described in Murienne et al. [10]. A two-camera stereovision system [10] imaged the specimen top-down (Fig. 7.1a, b), while a single-camera system acquired imaged of the specimen in profile (Fig. 7.1a, c). All three cameras were synchronized together and had a 0.025 mm/pixel image resolution. Sequential images were first acquired at baseline pressure 0.28 kPa, then the reference picture was taken at 5.41 kPa, and the deformed picture at 7.87 kPa. The high reference pressure 5.41 kPa was used to start from a bulged configuration of the membrane and allow for a uniform strain to develop under inflation to the maximum pressure 7.87 kPa.

7.2.2 3D-DIC Analysis

Vic-3D (Correlated Solutions, Columbia SC) was used to obtained the 3D reference positions and displacements on a 2D Cartesian grid with 0.125 mm spacing from the 3D system images. A sphere was fitted to the reference geometry (spherefit.m, Levente Hunyadi, 2010) and its radius R was used to create a reference spherical grid with 1° spacing in the meridional direction and 3° spacing in the circumferential direction. The displacements were interpolated onto that reference spherical grid to obtain (U_x, U_y, U_z) at each grid point (X, Y, Z). The horizontal displacement U_ξ in the ξ direction $\left(\xi = R\sin(\varphi)\right)$ was calculated from U_x, U_y as $U_\xi = U_x\cos(\theta) + U_y\sin(\theta)$. The vertical displacement was directly given by U_z. The new positions of a grid point n at each pressure step were defined as $x_n = X_n + U_{xn}$, $y_n = Y_n + U_{y_n}$, $z_n = Z_n + U_{z_n}$, and were used to calculate the undeformed $\left(L_{\phi_n}\right)$ and deformed $\left(l_{\phi_n}\right)$ grid lengths using central difference as:

$$L_{\varphi_n} = \sqrt{\left(X_{n+1} - X_{n-1}\right)^2 + \left(Y_{n+1} - Y_{n-1}\right)^2 + \left(Z_{n+1} - Z_{n-1}\right)^2},$$

$$l_{\varphi_n} = \sqrt{\left(x_{n+1} - x_{n-1}\right)^2 + \left(y_{n+1} - y_{n-1}\right)^2 + \left(z_{n+1} - z_{n-1}\right)^2} \tag{7.1}$$

from which the meridional stretch and strain could be obtained:

$$\lambda_{\varphi\varphi} = \frac{l_\varphi}{L_\varphi},\ E_{\varphi\varphi} = \frac{1}{2}\left(\lambda_{\varphi\varphi}{}^2 - 1\right) \tag{7.2}$$

The displacements and meridional strains were obtained for the 120 meridians of the reference spherical grid.

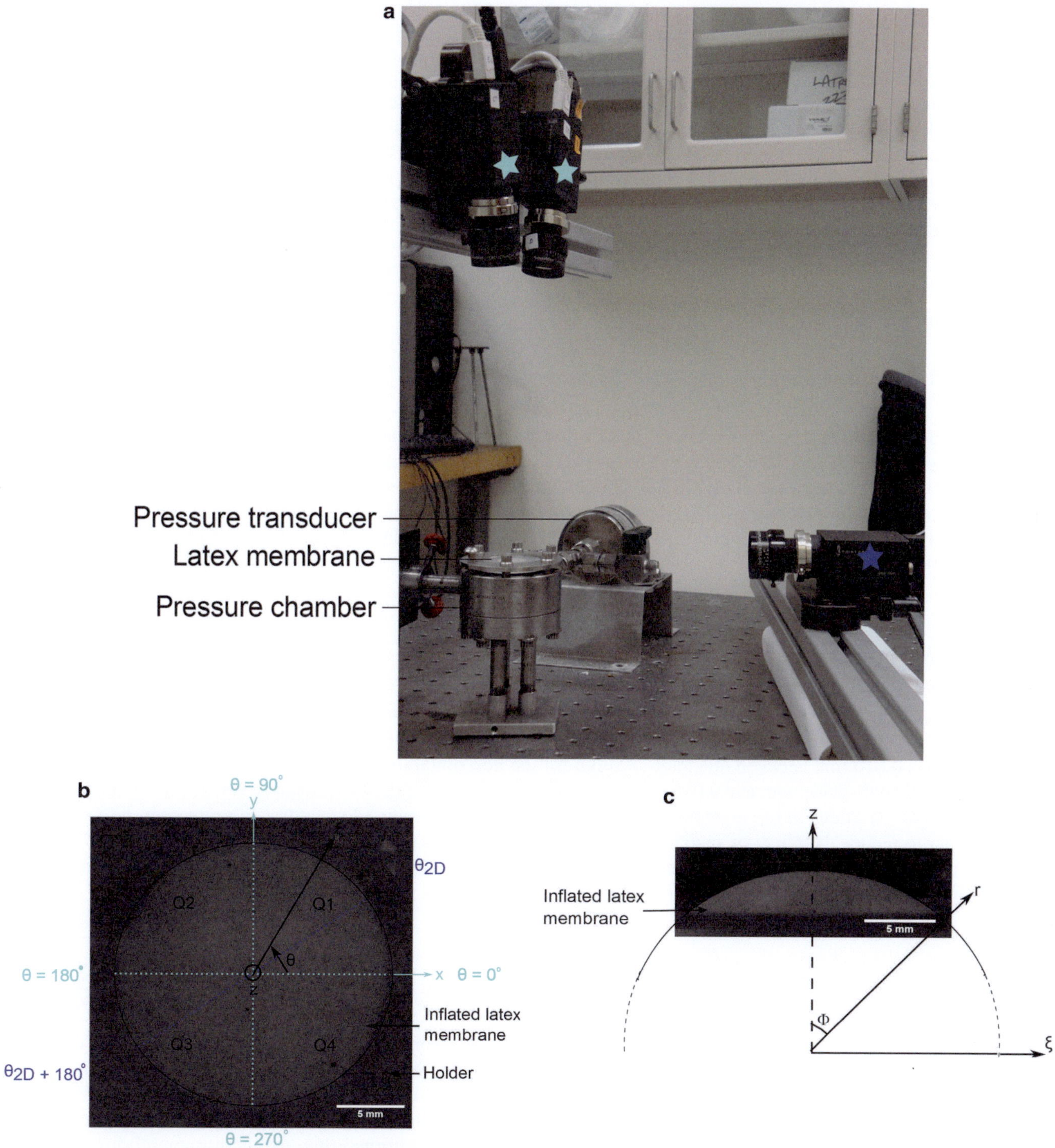

Fig. 7.1 (**a**) Setup picture showing the pressure chamber, latex membrane, pressure transducer, single-view (blue star) and stereovision (cyan stars) imaging systems. Top view (**b**) and side view (**c**) of the inflated latex membrane showing the projected radial direction ξ $\left(\xi = R\sin\left(\phi\right)\right)$ and vertical direction z. The cyan axis define the four quadrants used to analyze the 3D-DIC data and the blue axis define the 2 meridians used for the 2D-DIC data analysis

7.2.3 2D-DIC Analysis

Vic-2D (Correlated Solutions, Columbia SC) was used to extract the 2D reference positions ($X=\xi$, Z) as well as the horizontal displacements U_ξ and vertical displacements U_z at specific locations on the membrane boundary, from the 2D system images. Those locations corresponded to the ξ locations of the 3D reference positions on the spherical grid, starting from the center of the membrane which was obtained by fitting a circle to the membrane boundary (CircleFitByTaubin.m, Nikolai Chrenov, 2009). The strains were calculated from the displacements similarly to the 3D-DIC analysis but using the 2D version of the Eqs. (7.1) and (7.2). The displacements and strains were obtained along the 2 meridians captured on the images captured by the 2D imaging system (Fig. 7.1c).

7.3 Results

7.3.1 Static Conditions

Sequential pictures acquired at baseline pressure (taut membrane, no displacement) were analyzed to compare the static error and uncertainty of 3D-DIC and 2D-DIC in the horizontal and meridional displacements. The static uncertainty in the horizontal displacements was larger for 2D-DIC (0.0017 mm) than 3D-DIC (8.05e-4 mm), whereas the static uncertainty in the vertical displacements was larger for 3D-DIC (0.0032 mm) than 2D-DIC (8.63e-4 mm). The absolute static error in the horizontal displacements was 13 times smaller for 3D-DIC (3.33e-5 mm) than 2D-DIC (4.43e-4 mm), whereas the absolute static error in the vertical displacements was similar between both DIC methods.

7.3.2 Spatial Variations Across the Membrane

Displacements and strains obtained from 3D-DIC along the 120 meridians available showed large variations. When the displacements and strains were averaged over the 30 meridians of each quadrant, at each ξ position, the maximum difference along ξ was 0.016 mm between quadrants 2 and 4 for the horizontal displacement, 0.033 mm between quadrants 3 and 4 for the vertical displacement and 0.033 between quadrants 2 and 4 for the meridional strain. Those spatial variations were attributed to variations in the membrane thickness and texture as observed when transilluminating the latex membrane. In order to reduce the effect of spatial variations when comparing the outcomes from 2D-DIC and 3D-DIC in the rest of the study, we only used the 60 meridians of quadrants 1 and 3 from 3D-DIC as those quadrants contained the 2 meridians from 2D-DIC (Fig. 7.1b).

7.3.3 Displacements

For 3D-DIC, the variability across the meridians was larger for the vertical displacements than the horizontal displacements (Fig. 7.2a, b). The variability across the meridians was generally larger for the 3D-DIC displacement components than the 2D-DIC displacement components, but the difference was significant for the vertical displacements (Fig. 7.2a, b). Moreover, the absolute difference between the averaged 3D-DIC and 2D-DIC responses across all meridians was larger for the vertical displacements (0.014 mm on average over all ξ positions) than for the horizontal displacements (0.0024 mm on average over all ξ positions).

7.3.4 Meridional Strain

The variability in the meridional strains across the meridians was similar between 3D-DIC and 2D-DIC (Fig. 7.3). Assuming a constant meridional strain in the small region surrounding the apex of the membrane, the probability distribution for the 2D-DIC data (mean = 0.0303, SD = 0.0084) and the probability distribution for the 3D-DIC data (mean = 0.0274, SD = 0.0081) showed similar strain uncertainties.

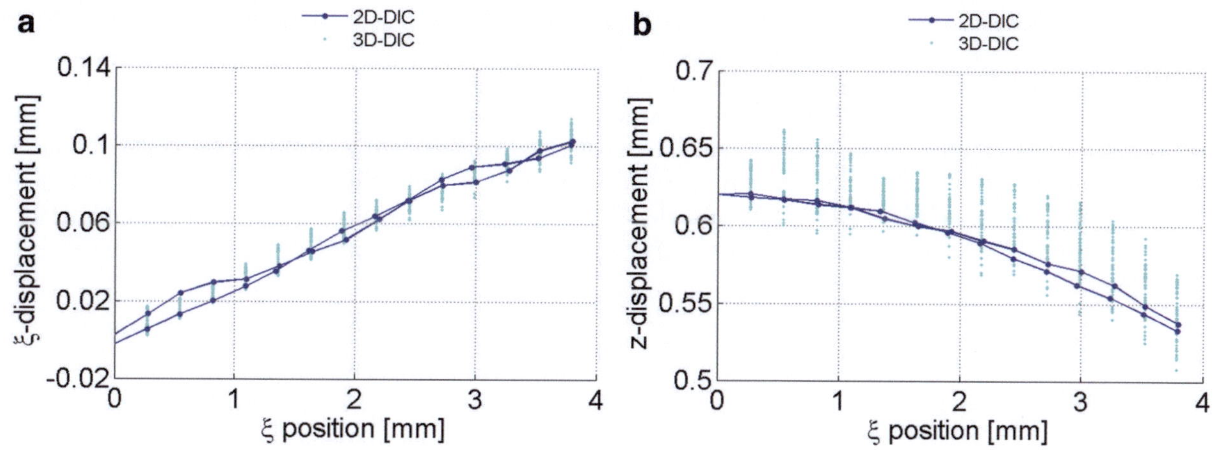

Fig. 7.2 Horizontal (**a**) and vertical (**b**) displacements along the ξ direction for the 60 meridians from 3D-DIC and the 2 meridians from 2D-DIC

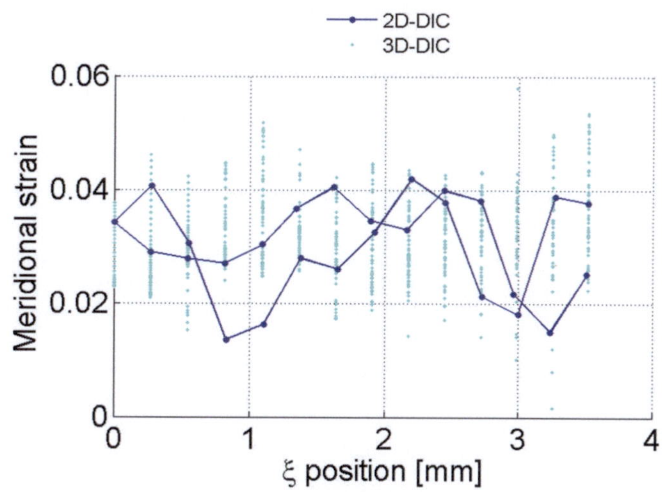

Fig. 7.3 Meridional strain along the ξ direction for the 60 meridians from 3D-DIC and the 2 meridians from 2D-DIC

7.4 Discussion

Under static conditions, 3D-DIC showed a smaller baseline absolute error and uncertainty in the horizontal displacements than 2D-DIC. This could be due to the correlation window containing some background and some speckle pattern at the membrane boundary for the 2D-DIC analysis. The baseline uncertainty in the vertical displacements was smaller for 2D-DIC than 3D-DIC, suggesting that 3D-DIC out-of-plane measures are more susceptible to noise, possibly due to the 3D reconstruction step.

Under inflation, the variability in displacements and strains was larger than under static conditions for both DIC methods. 3D-DIC showed a larger variability in the vertical displacements than horizontal displacements, suggesting an additional out-of-plane uncertainty in 3D-DIC, as reported in this study and by others [23, 24] under static conditions. 2D-DIC showed a smaller variability than 3D-DIC, especially in the vertical displacements. Moreover, the absolute difference between the averaged 3D-DIC and 2D-DIC data across the meridians was in the range of the 3D-DIC variability for both displacement components. The meridional strain uncertainty was similar between both DIC methods, which could be due to the interpolation of displacements for the 3D-DIC data analysis. However, the smoothing effect of interpolation was kept to a minimum by using an interpolation grid only twice as sparse as the array of data points. In addition, the absolute difference between the mean of the 3D-DIC and 2D-DIC strain distribution was within the range of the 3D-DIC strain uncertainty. In this study, the effect of the use of a non-telecentric lens for the image acquisition on the 2D-DIC outcomes was mitigated by the extraction of the displacements at the membrane boundary, which limited the effect of the membrane curvature. Moreover, the effect of defocusing during inflation was considered small, as the portion of the membrane contained in the correlation window remained within the depth of field of the 2D imaging system.

From the static and inflation results, 2D-DIC might be a valid option to study the profile behavior of stiff membranes, for which the vertical displacement is small, or small specimens under high magnification, for which the depth of focus is limited, under inflation. 2D-DIC might also be useful for specimens tested under controlled environments, which often limit or do not allow access for two-angled cameras. However, 3D-DIC should be used to fully capture the anisotropic response of heterogeneous materials to inflation, resulting from the material native microstructure or altered microstructure due to fabrication processes.

7.5 Conclusion

In conclusion, we performed the inflation of a latex membrane and compared the displacements and strains obtained from 3D-DIC and 2D-DIC on simultaneously acquired images. Our main findings were that under inflation, 2D-DIC had a smaller variability in displacements than 3D-DIC, especially for the vertical displacements. The absolute difference between the average 2D-DIC and 3D-DIC displacements was within the variability range of 3D-DIC for both displacement components. The strain uncertainty was similar between 2D-DIC and 3D-DIC and the absolute difference in mean strain between both methods was within the 3D-DIC strain uncertainty. Therefore, the use of 2D-DIC might be appropriate to study the profile behavior of a specimen under inflation when the experimental conditions do not encourage or allow the use of 3D-DIC.

References

1. Wang, C.C., Deng, J.M., Ateshian, G.A., Hung, C.T.: An automated approach for direct measurement of two-dimensional strain distributions within articular cartilage under unconfined compression. J. Biomech. Eng. **124**(5), 557–567 (2002)
2. Zhang, D., Eggleton, C.D., Arola, D.D.: Evaluating the mechanical behavior of arterial tissue using digital image correlation. Exp. Mech. **42**(4), 409–416 (2002)
3. Zhang, D., Arola, D.D.: Applications of digital image correlation to biological tissues. J. Biomed. Opt. **9**(4), 691–699 (2004)
4. Sutton, M., Ke, X., Lessner, S., Goldbach, M., Yost, M., Zhao, F., et al.: Strain field measurements on mouse carotid arteries using microscopic three-dimensional digital image correlation. J. Biomed Mater. Res. A **84**(1), 178–190 (2008)
5. Myers, K.M., Coudrillier, B., Boyce, B.L., Nguyen, T.D.: The inflation response of the posterior bovine sclera. Acta Biomater. **6**(11), 4327–4335 (2010)
6. Myers, K.M., Cone, F.E., Quigley, H.A., Gelman, S., Pease, M.E., Nguyen, T.D.: The *in vitro* inflation response of mouse sclera. Exp. Eye Res. **91**(6), 866–875 (2010)
7. Coudrillier, B., Tian, J., Alexander, S., Myers, K.M., Quigley, H.A., Nguyen, T.D.: Biomechanics of the human posterior sclera: age- and glaucoma-related changes measured using inflation testing. Invest. Ophthalmol. Vis. Sci. **53**(4), 1714–1728 (2012)
8. Soons, J., Lava, P., Debruyne, D., Dirckx, J.: Full-field optical deformation measurement in biomechanics: digital speckle pattern interferometry and 3D digital image correlation applied to bird beaks. J. Mech. Behav. Biomed. Mater. **14**, 186–191 (2012)
9. Tonge, T.K., Atlan, L.S., Voo, L.M., Nguyen, T.D.: Full-field bulge test for planar anisotropic tissues: part I–experimental methods applied to human skin tissue. Acta Biomater. **9**(4), 5913–5925 (2013)
10. Murienne, B.J., Jefferys, J.L., Quigley, H.A., Nguyen, T.D.: The effects of glycosaminoglycan degradation on the mechanical behavior of the posterior porcine sclera. Acta Biomater. **12**, 195–206 (2015)
11. Sutton, M., Yan, J., Tiwari, V., Schreier, H., Orteu, J.: The effect of out-of-plane motion on 2D and 3D digital image correlation measurements. Opt. Lasers Eng. **46**(10), 746–757 (2008)
12. Pan, B., Qian, K., Xie, H., Asundi, A.: Two-dimensional digital image correlation for in-plane displacement and strain measurement: a review. Meas. Sci. Technol. **20**(6), 062001 (2009)
13. Sutton, M.A., Yan, J., Deng, X., Cheng, C.S., Zavattieri, P.: Three-dimensional digital image correlation to quantify deformation and crack-opening displacement in ductile aluminum under mixed-mode I/III loading. Opt. Eng. **46**(5), 051003 (2007)
14. Becker, T., Splitthof, K., Siebert, T., Kletting, P.: Error estimations of 3D digital image correlation measurements. Proc. SPIE **6341**, 63410F (2006)
15. Siebert, T., Becker, T., Spiltthof, K., Neumann, I., Krupka, R.: Error estimations in digital image correlation technique. Appl. Mech. Mater. **7**, 265–270 (2007)
16. Girard, M.J., Suh, J.K.F., Bottlang, M., Burgoyne, C.F., Downs, J.C.: Scleral biomechanics in the aging monkey eye. Invest. Ophthalmol. Vis. Sci. **50**(11), 5226–5237 (2009)
17. Sokolowski, T., Gerke, K., Ahmetoglu, M., Altan, T.: Evaluation of tube formability and material characteristics: hydraulic bulge testing of tubes. J. Mater. Process. Technol. **98**(1), 34–40 (2000)
18. Vlassak, J., Nix, W.: A new bulge test technique for the determination of Young's modulus and Poisson's ratio of thin films. J. Mater. Res. **7**(12), 3242–3249 (1992)

19. Karimi, A., Shojaei, O., Kruml, T., Martin, J.: Characterisation of TiN thin films using the bulge test and the nanoindentation technique. Thin Solid Films **308**, 334–339 (1997)
20. Xiang Y, Chen X, Vlassak JJ. The mechanical properties of electroplated Cu thin films measured by means of the bulge test technique. MRS Proc. 695, L4–9. Cambridge Univ. Press (2001)
21. Huang, C., Lou, W., Tsai, C., Wu, T.C., Lin, H.Y.: Mechanical properties of polymer thin film measured by the bulge test. Thin Solid Films **515**(18), 7222–7226 (2007)
22. Berdova, M., Ylitalo, T., Kassamakov, I., Heino, J., Törmä, P.T., Kilpi, L., et al.: Mechanical assessment of suspended ALD thin films by bulge and shaft-loading techniques. Acta Mater. **66**, 370–377 (2014)
23. Hu, Z., Xie, H., Lu, J., Wang, H., Zhu, J.: Error evaluation technique for three-dimensional digital image correlation. Appl. Opt. **50**(33), 6239–6247 (2011)
24. Ke, X.D., Schreier, H., Sutton, M., Wang, Y.: Error assessment in stereo-based deformation measurements. Exp. Mech. **51**(4), 423–441 (2011)

Chapter 8
Determination of Cardiac Wall Deformations from MRI Images

Cesar Augusto Sciammarella, Luciano Lamberti, and Antonio Boccaccio

Abstract Three-dimensional deformation analysis of human organs is very important from both diagnostic and therapeutic point of view. For example, comparing the deformation field in healthy and pathologic cardiac walls in the systolic phase allows to gather early and accurate information on the onset of heart diseases. MRI tagging is utilized in medicine to visualize with a great deal of detail the structure and morphology of tissues. The tagging process introduces a volumetric system of planes of reference similar to the process of introducing a grating in the 3-D moiré method in transparent media. The paper will analyze the kinematics of 3-D deformation fields and the fundamental concepts involved in 3-D deformation analysis within the restrictions imposed by the MRI method thus providing solutions for the inherent shortcomings encountered in the MRI tagging technique.

Keywords MRI tagging • Digital moiré • Large deformation analysis • Rotations • Deformation of cardiac wall

8.1 Introduction

The authors, in a previous publication [1], presented the basic foundations of the MRI technique to detect deformations in human organs through the tagging process. Reference [1] summarizes the fundamental physics of MR and the methodology of MRI to produce images. In the MRI methodology, the magnetic vector **B** plays the role of the illuminating vector field **E** in optical methods since it can penetrate the human body different tissues and generate signals that reflect the state of magnetization of a voxel located inside a tissue. MRI through the tagging process makes it possible to measure geometrical changes of human tissues. Continuum Mechanics principles can be applied to human organs. The tagging process introduces a volumetric system of planes of reference similar to the process of introducing a grating in the 3-D moiré method [2, 3] in transparent media. This is done by manipulating the state of magnetization of the voxels, resulting in periodic intensity changes that are attached to the voxels. However, these changes are temporary and fade away with time creating problems with information acquisition timing. Tagging has an effect in the quality of the produced images. Furthermore, the MRI method experiences a degradation of image quality caused by cardiac motions and respiratory motions of the observed individual. There is an extensive literature covering different solutions for the before mentioned problem, a good review up the time of publication can be found in [4]. One of the more promising current approaches to the solution of many of the problems of MRI is the compressing sensing technology (CS). This recent research field concentrates on operations performed in the frequency space rather than in the physical space [5]. The basic idea behind CS is to recover signals and images from a reduced number of measurements. The application of CS cuts down the time of scan required by the current prevalent MRI methods thus increasing the capability of MRI to provide information within the limitations posed by the physics of MR. Transform-based compression of signals is an extensively utilized process in JPEG, JPEG-2000 and MPEG standards, commonly used methods of compression for digital images, for example in digital photography.

The objective of the current paper is to analyze the kinematics of 3-D deformation fields and the fundamental concepts involved in 3-D deformation analysis within the restrictions imposed by the MRI method thus providing solutions for the above mentioned basic shortcomings encountered in the MRI tagging technique.

C.A. Sciammarella
Department of Mechanical, Materials and Aerospace Engineering, Illinois Institute of Technology, 10 SW 32 St., 60616 Chicago, IL, USA

L. Lamberti (✉) • A. Boccaccio
Dipartimento di Meccanica, Matematica e Management, Politecnico di Bari, Viale Japigia 182, 70126 Bari, Italy
e-mail: luciano.lamberti@poliba.it

S.A. Tekalur et al. (eds.), *Mechanics of Biological Systems and Materials, Volume 6*,
Conference Proceedings of the Society for Experimental Mechanics Series, DOI 10.1007/978-3-319-21455-9_8

8.2 Deformations Data Analysis

We start with some fundamental elements of 3-D kinematics of deformations. In [6], a comprehensive presentation of the different methods that have evolved to detect and analyze displacements in the continuum and comparisons among these different methods is presented including HARP, one of the most applied method in the study of deformations of human or animal organs. The current introduction emphasizes the role of signal analysis methods that were developed in the more general context of signal analysis.

One of the starting points of the evolved technology is the utilization of geometrical spaces other than the physical space of rational mechanics. There are considerable advantages to operate in spaces other than the physical space as clearly indicated in [5, 6].

Figure 8.1 shows the physical space with the corresponding carrier planes utilized to detect displacements. At the bottom left of Fig. 8.1b it is shown a descriptive geometry representation of the 3-D space. The traces of the planes are indicated by green lines. If the volumetric tagging signals are sinusoidal the traces would be sinusoidal. However, in the actual process of tagging not only one harmonic is generated but many harmonics are produced and the traces become thick lines. Figure 8.1a shows an actual tagged image of the heart showing the traces of the tagging planes in cross sections of the heart. The volumetric physical space has it corresponding volumetric frequency space. The relationship between the two spaces is:

$$p_o = \frac{1}{k_o} \tag{8.1}$$

Furthermore, the correspondence between the two spaces is shown in Fig. 8.2 that represents the families of tagging planes corresponding to each one of the coordinates planes. In the frequency space each family of planes is represented by a dot. Since each tagging family contains many harmonics, the corresponding order will be present in the frequency space; this particular representation corresponds to the Fourier transform (FT). There are several points that need to be stressed. The preceding representation deals with the reference image. If the medium is deformed then the dot representing the harmonic becomes a distribution of intensities around the original frequency. Points in the frequency space contain information corresponding to the entire field of view in the physical space. Hence, points in the frequency space inside a given cube of sides k_o contain information concerning the overall volume under analysis. Since the tagging planes are moving with the tissues where they are tagged to, rigid body translations will not influence the relative displacements provided by the tagging planes. However, rigid body rotations will influence the results since they will change the spatial frequencies of the projected points. Since one of the main purposes of recording cardiac images is an evaluation of the heart as a blood pumping device and this evaluation is connected to relative displacements and deformations measured by rigid body motion invariant strain tensors,

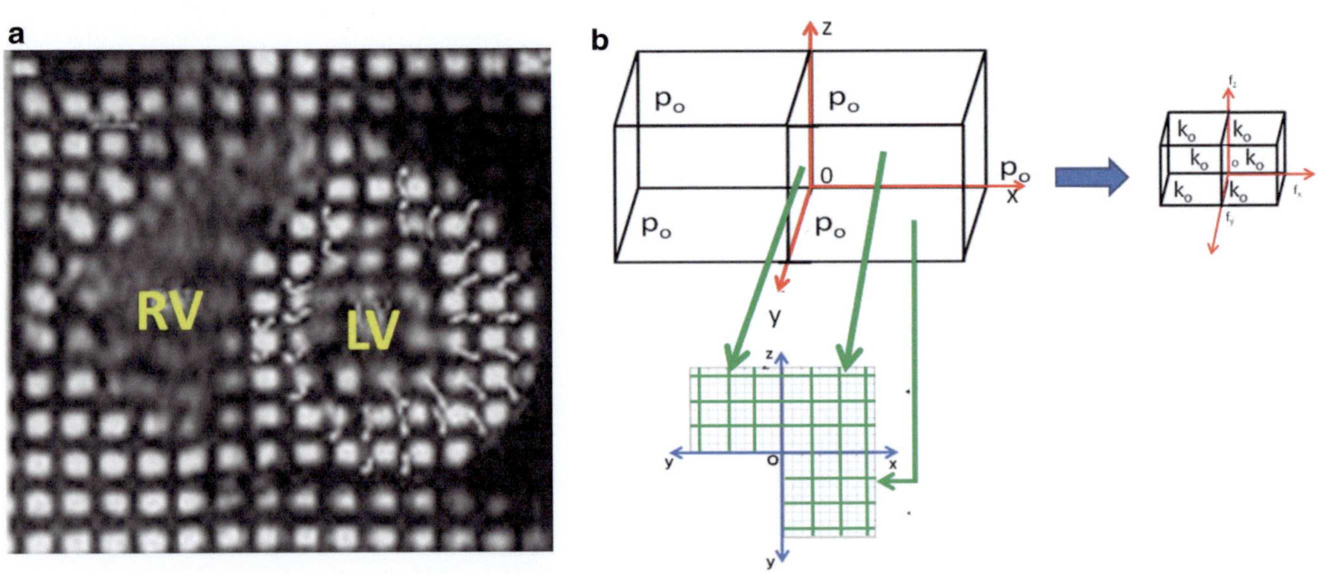

Fig. 8.1 (**a**) Volumetric 3-D planes system for determination of relative displacements and the corresponding frequency space volume of tagged heart LV and RV (plane x–y). (**b**) Descriptive geometry representation of the 3-D space

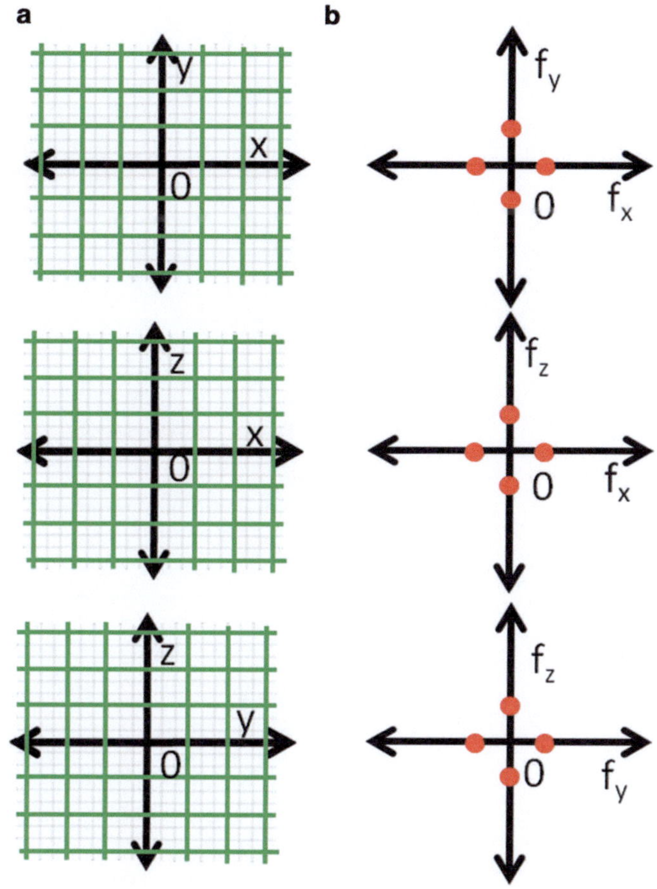

Fig. 8.2 (**a**) Tagging planes in the three coordinate planes of a Cartesian system. (**b**) Frequency *space dots* represent the families of planes in the direction perpendicular to the *lines* that join the *dots*

let us first consider this particular aspect of the problem. In the following, analysis large deformations will be considered and the variable time will be introduced as part of the recording of a sequence of images that follow the deformation of a body.

Figure 8.3 represents a vessel in which the thicknesses of the walls are smaller than the overall dimensions of the vessel itself.

It can be imagined that the volume of the vessel has been tagged with families of planes that cover the entire volume. The recordings of the changes in geometry of the vessel are done in successive times to, t_1, t_2, t_3,… so that the time history of the vessel can be followed. It will be assumed that the rules of selection of tagging planes are satisfying the spatial frequencies requirements given in [7] and that time recordings are such that the rules of dynamic recording of images are satisfied [7] and also the timing is such that the selected intervals provide the desired information, for example, about the volume changes of the vessel. Provided these conditions are fulfilled, it is possible to relate the changes of the vessel configuration by applying the rules of continuum mechanics and the developed theory of analysis of deformations utilizing deterministic signals known in the literature as digital moiré [7]. All the necessary operations will be performed utilizing Cartesian coordinates $\mathbf{x}_i(i=1, 2, 3)$ and the corresponding projecting displacement components will be called $\mathbf{u}_i$. In order to make the applied steps as clear as possible, vectorial notation will be utilized in most of derivations. The ordinary rules of Cartesian tensors will be applied whenever convenient to reduce the equations extent. In [6], it was pointed out an important distinction concerning displacements, distinguishing absolute displacements with respect to a chosen global spatial frame of reference, and relative displacements of the points of the continuum where the selection of the zero relative displacement planes are selected according to convenience in the development of the analysis.

Whenever we will talk about displacements, these will be relative displacements unless it is explicitly indicated. Let us first consider two successive images, for example t_o and t_1 in Fig. 8.3. The planes of relative displacement zero have been selected in coincidence with the Cartesian axes of symmetry of the vessel 0–x_i, and the considered volumes will be those corresponding to the time $t=0$ and $t=t_1$ in the cross-sections of the short axes, that we will assume correspond to $x_3=a$, where a is the distance of the considered section to the zero-coordinate plane of reference. We need to consider the information in the direction of the long axis of the vessel, the x_3 axis, and according to Fig. 8.3 we are given information along the shortest axis of the cross section of the vessel 0–x_1. Points that have the same orders with respect to the zero orders planes according

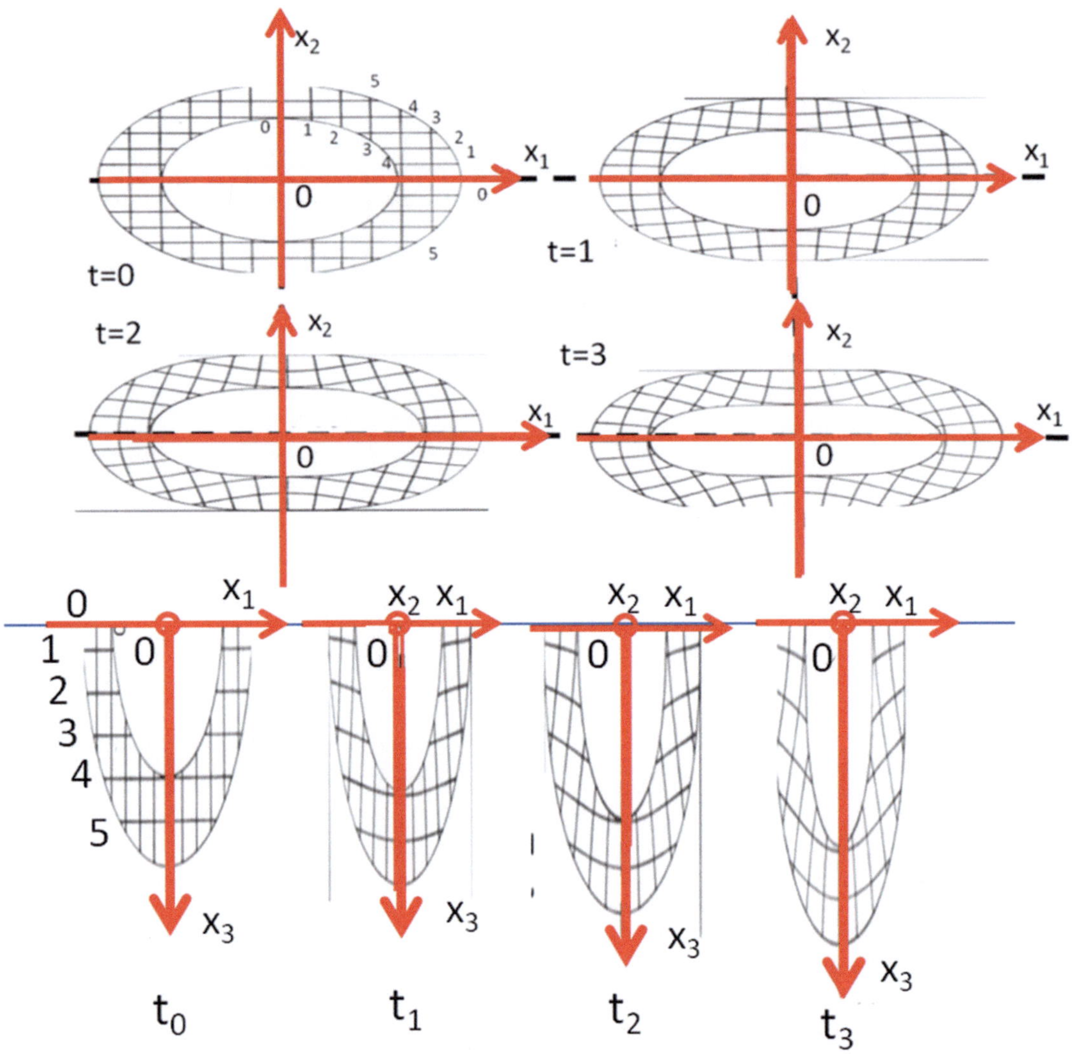

Fig. 8.3 Cross-sections of successive images in time of a tagged vessel that undergoes large deformations preserving axis of symmetry

with the number assigned in Fig. 8.3 in the projection planes correspond to the same material point in the initial position P that after relative displacement ends up in the position Q (see Fig. 8.4).

For the plane x_1x_2 of Fig. 8.4 it holds,

$$\mathbf{u}_{12} = u_1 e_1 + u_2 e_2 \tag{8.2}$$

where the $\mathbf{e}_i$ are the corresponding versors and u_1 and u_2 are the components obtained from the carrier gratings analysis.

For the plane x_1x_3 of Fig. 8.4 it holds,

$$\mathbf{u}_{13} = u_1 e_1 + u_3 e_3 \tag{8.3}$$

where u_1 and u_3 are obtained from the carrier grating analysis.

Finally, for the plane x_2x_3 the corresponding relationships are,

$$\mathbf{u}_{23} = u_2 e_2 + u_3 e_3 \tag{8.4}$$

From Eqs. (8.2–8.4) one obtains,

$$\mathbf{u} = u_1 e_1 + u_2 e_2 + u_3 e_3 \tag{8.5}$$

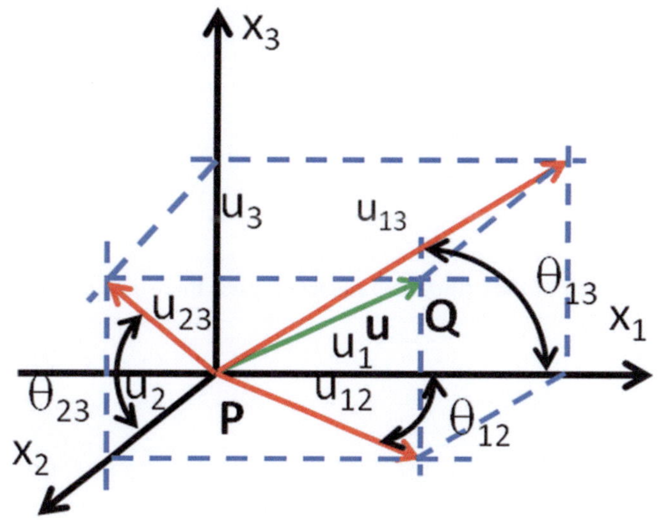

Fig. 8.4 Points P and Q correspond to the same material point because they have the same relative orders at $t=t_o$ and $t=t_1$ in the corresponding volumes

An important conclusion can be drawn: with two families of tagging planes in the cross-sections x_1x_2 corresponding to the short axes of the vessel and one family of planes in the longitudinal cross section providing u_3 it is all the information required to get the relative displacements between a reference configuration and a final configuration.

The length of the displacement vector is:

$$|u_r| = \sqrt{u_1^2 + u_2^2 + u_3^2} \tag{8.6}$$

The angles made by the displacement vectors with the coordinate axes can be computed as,

$$\theta_{12} = arctg\frac{u_2}{u_1} \tag{8.7}$$

and the corresponding similar equations for the other components.

The relative displacement field is governed by a system of differential equations,

$$\frac{dx_2}{dx_1} = \frac{u_2(x_1, x_2, x_3)}{u_1(x_1, x_2, x_3)} \tag{8.8}$$

$$\frac{dx_3}{dx_1} = \frac{u_3(x_1, x_2, x_3)}{u_1(x_1, x_2, x_3)} \tag{8.9}$$

The above differential equations define field trajectories that were called in [6] trochias. The trochias are perpendicular to the lines of equal inclination, given by Eq. (8.7) and the other corresponding equations, which can be called trochias isoclinics to distinguish them from the isoclinics defining directions of principal stresses.

There is a system of lines of equal displacement, called isokinetics by Durelli [8], that can be obtained from the isothetic lines through Eq. (8.6). The above mentioned families of lines describe the displacement field of the 3-D continuum.

8.3 Computation of the Derivatives of the Displacements

The whole process of the computation of the derivatives was analyzed in detail in [6]. In the present case, the only difference is that, because of the limitation of MRI, the family of planes orthogonal to the planes whose normal has the direction of the axis x_3 is missing. Since we are dealing with digital moiré, we obtain the Eulerian configuration of the analyzed body. Hence, to obtain the derivatives it is necessary to compute the tensor J defined in [6],

$$J = \begin{bmatrix} \frac{\partial u_1}{\partial x_1} & \frac{\partial u_1}{\partial x_2} & \frac{\partial u_1}{\partial x_3} \\ \frac{\partial u_2}{\partial x_1} & \frac{\partial u_2}{\partial x_2} & \frac{\partial u_2}{\partial x_3} \\ \frac{\partial u_3}{\partial x_1} & \frac{\partial u_3}{\partial x_2} & \frac{\partial u_3}{\partial x_3} \end{bmatrix} \tag{8.10}$$

The lack of the orthogonal planes implies that some components of the J tensor are unknown,

$$J = \begin{bmatrix} \frac{\partial u_1}{\partial x_1} & \frac{\partial u_1}{\partial x_2} & X \\ \frac{\partial u_2}{\partial x_1} & \frac{\partial u_2}{\partial x_2} & X \\ X & X & \frac{\partial u_3}{\partial x_3} \end{bmatrix} \tag{8.11}$$

However, it was pointed out before that there is enough information in the patterns to get the complete displacement vector. In Sect. 6.1 of [6], a relationship between derivatives in differential geometry was derived. Applying Eq. (21) of [6] to plane x_1x_3 of Fig. 8.4, it can be written

$$\frac{\partial u_3}{\partial x_1} = \left(1 + \frac{\partial u_1}{\partial x_1}\right) \tan\theta_{13} \tag{8.12}$$

A similar equation can be obtained for the plane x_2–x_3,

$$\frac{\partial u_3}{\partial x_2} = \left(1 + \frac{\partial u_2}{\partial x_2}\right) \tan\theta_{23} \tag{8.13}$$

Utilizing the expressions corresponding to the orthogonal directions, it follows

$$\frac{\partial u_1}{\partial x_3} = \left(1 + \frac{\partial u_3}{\partial x_3}\right) \tan\theta_{31} \tag{8.14}$$

$$\frac{\partial u_2}{\partial x_3} = \left(1 + \frac{\partial u_3}{\partial x_3}\right) \tan\theta_{32} \tag{8.15}$$

Utilizing Eqs. (8.12–8.15) it is possible to complete at a point the **J** tensor. By applying this tensor one can select the desired strain tensor to be utilized to complete the kinematic analysis.

8.4 Generalization of the Procedures for Digital Moiré Carrier Signal Analysis

In the preceding sections were outlined the different steps required for the analysis of carrier fringes experiencing large deformations but preserving the symmetry axes. The process can now be generalized to less restrictive conditions. It becomes clear that the order number of the processed signals is an essential element to follow the deforming medium. In the case of carrier signals the order determination is a straightforward procedure because the orders can be determined unambiguously in the frequency space (see Chap. 8 of [7]). As shown in Sect. 14.4 of [7], from the orders in the FT space of a given plane it is possible not only to get the in-plane components but also the out-of-plane components by manipulating the filtering procedures assuming that out-of-plane motions are small quantities. Consequently, it is possible to connect the deformed configuration at any stage of the deformation with the reference configuration and compute relative displacements between two given configurations.

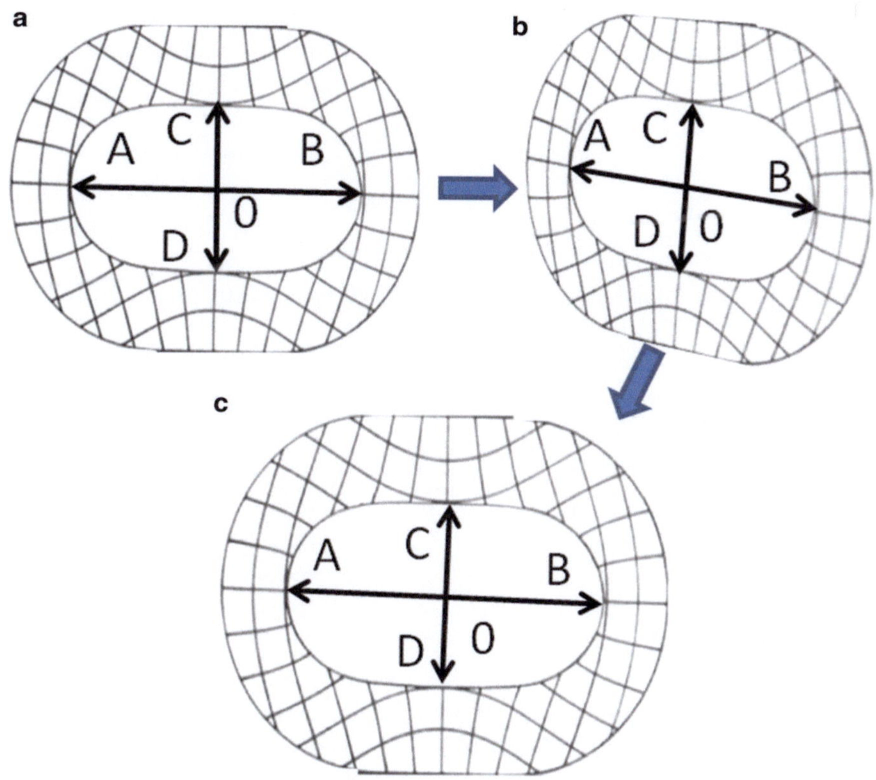

Fig. 8.5 (**a**) Image of vessel (reference position). (**b**) Image after vessel has experienced a change of shape plus an arbitrary rigid body motion in 3-D. (**c**) The rigid body motion has been removed and the change of shape is recovered

In the case illustrated in Fig. 8.3, an implicit assumption is that upon deformation of the analyzed vessel neither out-of-plane nor in-plane rigid rotations have taken place between the captured cross-sections images. In what follows the effect of relative rotations between sections will be considered.

Since the images of MRI are produced by slices, the amounts of the out-of-plane rotations that need to be considered are limited to the thickness of a slice. This is not the case for in-plane rotations.

Figure 8.5a shows the reference position of a vessel that has been tagged with a family of orthogonal planes. It can be seen the intersection of the reference planes with the image plane. In the subsequent image, Fig. 8.5b, the vessel has experienced a deformation plus a rigid body motion in 3-D. One can see that the intersection of the tagging planes with the image plane identify the locations of the points in the reference and final positions (i.e. A, B, C, D). However, the shape observed in the plane is a consequence of the combined effects of the deformation and rigid body motion. In Fig. 8.5c, the effect of the rigid body motions has been removed and the vessel deformation can be analyzed as it has been described in the preceding sections. As it has been mentioned before, translations will not affect the orders of the carriers and only rigid body rotations will alter the carrier frequencies because of the changes in geometry of the observed image. In the preceding argument we assume that the translations are limited in value in the sense that they will not produce changes on the scale of the image. Translations can be removed by a scale correction factor that can be included as an additional parameter in the processing of data. We will concentrate our attention in the effect of rigid body rotations and in how corrections can be performed within the previously mentioned limits.

8.5 Removal of the Effect of Rigid Body Rotations

In each of the coordinate planes the rotation is characterized by one angle: for example, in the x_1x_2 plane,

$$x_{1R} = \cos\alpha_1 x_1 + \sin\alpha_1 x_2 \tag{8.16}$$

$$x_{2R} = -\sin\alpha_1 x_1 + \cos\alpha_1 x_2 \tag{8.17}$$

Since there are three coordinate planes, we have three independent angles α_1, α_2 and α_3 that should be known in order to perform the rotation correction. The corresponding rotation with respect to axis x_1 is given by,

$$\begin{bmatrix} x_{1R} \\ x_{2R} \\ x_{3R} \\ 1 \end{bmatrix} = \begin{bmatrix} 1 & 0 & 0 & 0 \\ 0 & \cos\alpha_1 & \sin\alpha_1 & 0 \\ 0 & -\sin\alpha_1 & \cos\alpha_2 & 0 \\ 0 & 0 & 0 & 1 \end{bmatrix} \begin{bmatrix} x_1 \\ x_2 \\ x_3 \\ 1 \end{bmatrix} \tag{8.18}$$

For the rotations around the other two axes the corresponding rotation matrices are:

$$[R_2] = \begin{bmatrix} \cos\alpha_2 & 0 & \sin\alpha_2 & 0 \\ 0 & 1 & 0 & 0 \\ -\sin\alpha_2 & 0 & \cos\alpha_2 & 0 \\ 0 & 0 & 0 & 1 \end{bmatrix} \tag{8.19}$$

$$[R_3] = \begin{bmatrix} \cos\alpha_3 & \sin\alpha_3 & 0 & 0 \\ -\sin\alpha_3 & \cos\alpha_3 & 0 & 0 \\ 0 & 0 & 1 & 0 \\ 0 & 0 & 0 & 1 \end{bmatrix} \tag{8.20}$$

The above equations provide the changes of coordinates that every point in the tagged volume experiences as the object imaged deforms and experiences rigid body rotations. Let us consider for example point A in Fig. 8.5. For this point one needs to know the values of x_i and the values of α_i. The following equation relates the coordinates of the point $\mathbf{x}$ and the rotated coordinates $\mathbf{x}_R$:

$$X^R = [R]x \tag{8.21}$$

with $[R]=[R_1][R_2][R_3]$.

In this way one can determine the "rotated" coordinates x_{1A} and x_{2A} and locate A. In order to locate additional points along the horizontal axis towards the outside of the vessel, one must remember that information is in discrete voxels. The number of voxels along the corresponding lines in Figs. 8.5b and Fig. 8.5c is different: there are less voxels in (b) than in (c). Consequently, applying a simple rotation, "holes" will appear in (c). To solve this problem it is not enough to apply a rotation transformation, interpolation methods must be introduced. There are routines available to achieve this objective. For example, in MATLAB file Exchange: 3D rotation of a 3D image performs the operations required to go from the image of Fig. 8.5b to the image of Fig. 8.5c.

8.6 Determination of the Rigid Body Rotations

The process described in the preceding section assumes that one has the knowledge of the rigid body rotations that have taken place between successive images. These quantities are unknown in general and the problem has an additional difficulty because the body, in the current case, the heart, moves at the same time that deforms. The solution of the problem can be found in the invariance of the 2-D FT upon rotation. When a given image is rotated in its plane, the FT is rotated of the same amount. Hence, the procedure to obtain the rigid body rotation along the three reference axes is to locate features of the image in the reference state and in the deformed state that are in correspondence through a rigid body rotation. This task is facilitated by the fact that the images are covered by a grating system and through the positions of the diffraction spots it is possible to get the amount of rotation experienced by each one of the projections of an element of volume in the reference planes.

Figure 8.6 shows the spectrum of the tagging planes in the reference state. There is a number of harmonics corresponding to each family of planes. It will be assumed that the deformations of the analyzed body although large do not highly modulate the carrier fringes. Each of the three planes (i, j) is analyzed. The body has experienced an arbitrary rigid body rotation in 3-D. A rotation around the axis perpendicular to the plane x_i–x_j will rotate the spectrum and will expand the signals to harmonics around the carrier that will correspond to the maximum intensity in the frequency space. As it can be observed in Fig. 8.6, the carrier frequencies after the arbitrary spatial rotation are no longer along the rotated axes; the changes in the

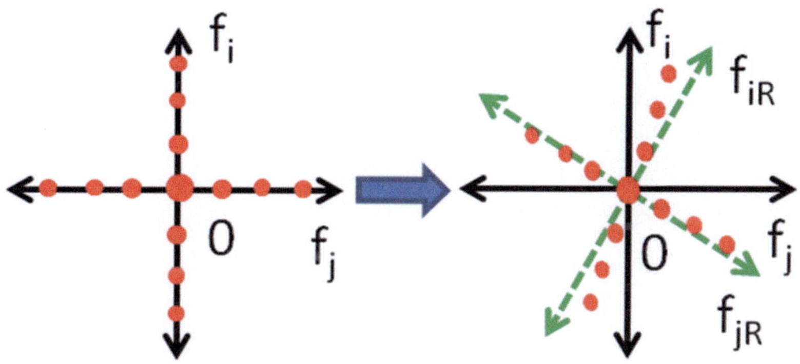

Fig. 8.6 Spectrum of the harmonics corresponding to the intersection of the tagging planes with the plane under analysis

position are due to the effect of the rotations along the other two axes that have been seen to modify the geometry of the sections of the body in the analyzed plane.

In summary, in order to get rotations around the three reference axes, it is necessary to determine three parameters that describe the rigid body motion matrix of Eq. (8.21). In the literature of image transformation the involved operation is called image registration. Then are needed three parameters that best match the reference and the deformed images to apply the processes described in Sects. 8.2 and 8.3. In the language of optimization, it must be assumed a cost function that can be minimized with an optimization algorithm. In the present case, we have to minimize the distances in the frequency space between corresponding orders caused by the actual rotation of the axis and the position of the same orders in the actual recorded images. This operation has to be performed with respect to the rotation around the corresponding axes as described in Sect. 8.5 but in place of performing the computations in the physical space the computations are done in the frequency space. The transference of the necessary operations from the physical space to the frequency space minimizes the number of points needed in the search of a solution.

The position of an order in the reference image is given by,

$$f_n = f_{1n} e_{f1} \tag{8.22}$$

In the deformed image,

$$f_{n1R} = \cos\alpha_1 f_{1n} \tag{8.23}$$

$$f_{n2R} = -\sin\alpha_1 f_{1n} \tag{8.24}$$

where α_1 is an assumed value of the angle, for example the angle corresponding to the first order.

The same order will appear in a different positions in the frequency space (see Fig. 8.7) due to the general rotation experienced by the observed body. For example, for the third order spot in Fig 8.7 ($i \neq j$), it can be written

$$n_3 = f_{i3} e_{i3} + f_{j3} e_{j3} \tag{8.25}$$

The difference of the positions of the spots hence is,

$$n_{3R} - n = (f_{i3R} - f_{i3}) e_{i3} + (f_{j3R} - f_{j3}) e_{j3} \tag{8.26}$$

Three equations can be written, one for each of the frequency planes x_i–x_j. These equations can be written for several orders. The next step is to obtain the optimum values of the α_i that minimize the distance between the position of the corresponding orders in each of the reference planes and for each of the considered orders,

$$d_{nij} = \sqrt{(f_{i3R} - f_{j3})^2 + (f_{j3R} - f_{j3})^2} \tag{8.27}$$

One can make an initial parameter estimate by merging one of the available orders and by iteration. The iteration is finished when, for example, the value of the difference has reached a value that is within the limits of accuracy that can be reached in view of the images quality. Once the values of the α_i are computed, the methodology explained in Sect. 8.5 is applied.

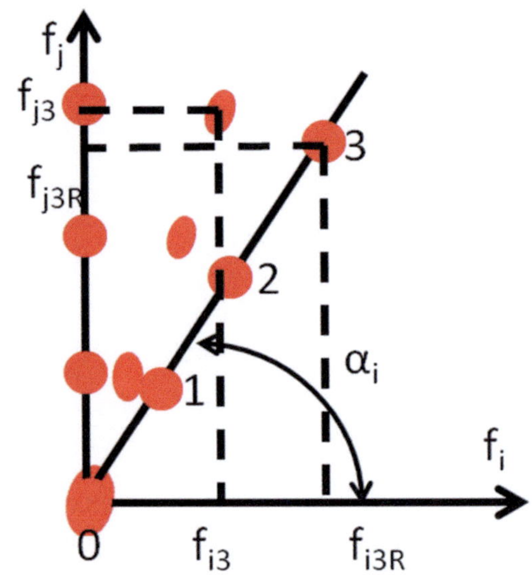

Fig. 8.7 Position of the orders of diffraction associated with a given x_i–x_j tagging plane upon a rigid body rotation around an axis perpendicular to the plane. Position of the same orders due to the rotated image (Fig. 8.5b)

8.7 Implementation of the Developed Methodology in MRI Images

After the corrections introduced by rigid body rotations the task of data processing is not complete, it remains a number of additional sources of image distortion. The images produced by the MRI system depend on the software developed by a particular manufacturer of the system and artifacts that appear in the image are not easy to correct. There are also limitations that arise from the method itself that complicate the task of correcting artifacts in an image. The MR method obtains spatial information by superimposing magnetic fields to a basic magnetic field that "illuminates" the observed volume, these magnetic fields produce gradients G_i that define the positions of points in the observed volume [1]. The magnetization vector takes a time called in the literature T_1 to build up and the time T^*_2 to decay and there is another parameter of importance T_{rec} the recording time: the interaction among these parameters in the generation of an image is discussed in [1]. A more complete discussion of the factors influencing the image quality can be found in [9]. One way of solving some of these problems is to perform what is called phantom experiments. This means to introduce an object that is independently analyzed and then obtaining records in the MRI system. The comparison of the independently obtained results with the MRI results makes it possible to develop correction algorithms.

8.8 Application of Digital Moiré to the Analysis of MRI Images of the Heart

The application of digital moiré to heart images was illustrated with some examples in [1, 6, 10]. It was shown that the traditional HARP approach and digital moiré are in good agreement in spite of all the different problems that have been pointed out. The necessary steps will now be covered in some more detail emphasizing the role of digital moiré in providing solutions to some of the artifact problems.

In order to understand the required steps in the analysis of the heart operation, it is necessary to have knowledge of the heart anatomy (see Fig. 8.8). The heart is a complex pumping system with two main vessels right ventricle (RV) and left ventricle (LV) with a common wall called the septum. At the upper end the two ventricles are connected by valves that control the inflow and outflow of blood regulated by electric signals that control the operation of the heart, ekg. As shown in the cross-section of the short axis (Fig. 8.8), in each stage of the heart cycle the walls contract or expand.

Figure 8.9 shows the tagged images of the RV and LV in two subsequent times [11]. The necessary steps to retrieve deformation information from the reference state to the deformed state are now outlined. The MRI tagging makes it possible to follow the heart wall changes at the different times of the cycle. The ultimate end of this process is to extract clinically useful information to evaluate the heart performance of an individual and compare normal heart functioning with different pathologies of the heart. The corrections caused by rigid body rotations were already discussed. It is necessary to see how additional procedures can provide corrections to artifacts introduced in the images caused by the many other factors indicated previously. We aim to connect these operations to the regular process applied to the analysis of fringe data in digital moiré. The two ventricles form a thin wall vessel that encloses two cavities.

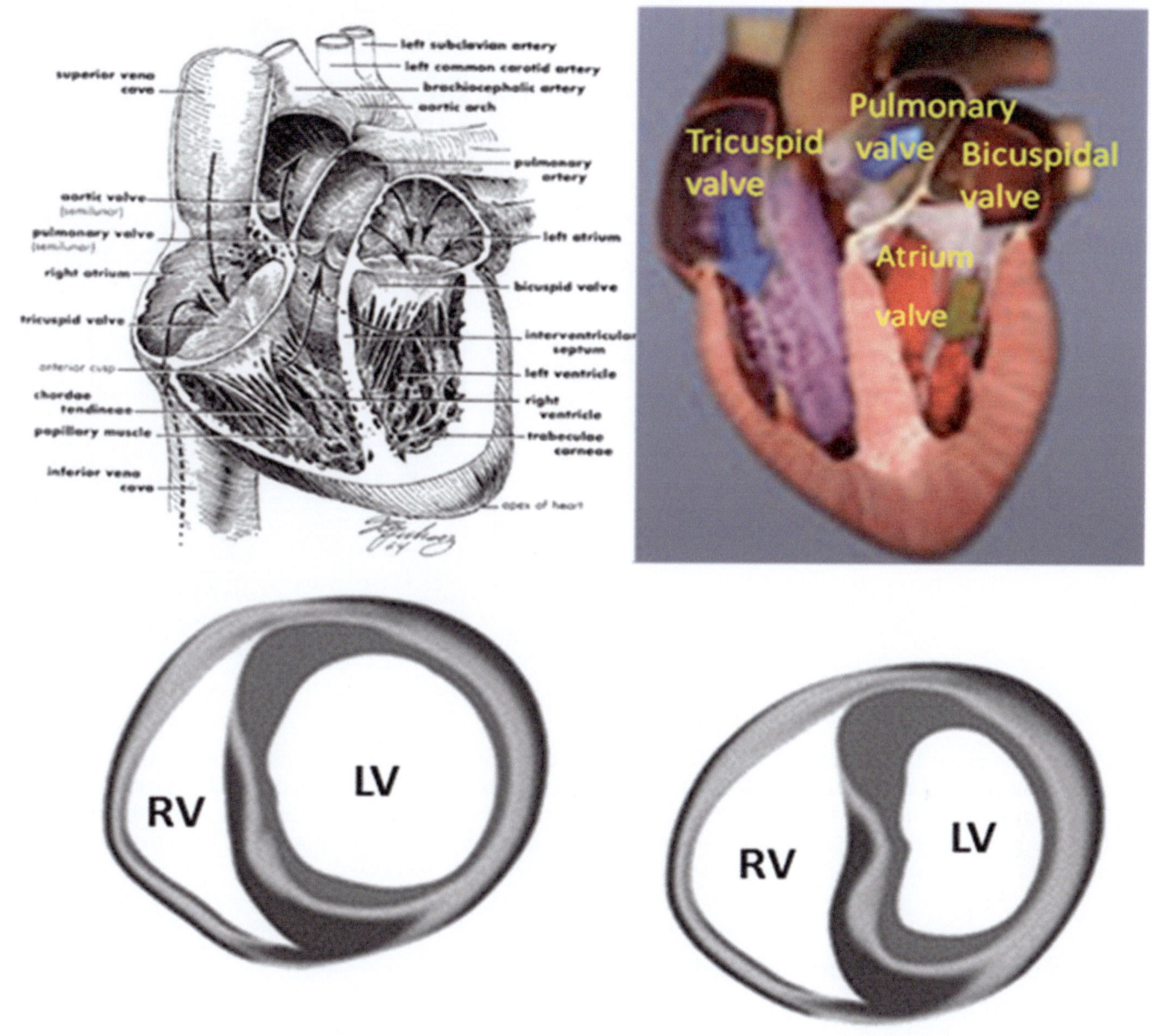

Fig. 8.8 Heart-long axis cross section (*top*) and short-axes cross sections (*bottom*)

A main problem in the fringe information retrieval is the effect of field discontinuities that appear at the boundaries of the region being analyzed. In the present case, fringe pattern analysis was done in the FT domain and in order to increase the speed of the operations the Fast Fourier Transform (FFT) method was applied. The utilized filters were symmetric and discontinuities at the boundaries could produce erroneous results propagating inside the region of interest. One can generate adaptive filters that modify their structure as the scanning of the image advances from the boundary of the domain where the fringes are defined to the interior of the domain. However, this approach is computationally more intensive that the solution adopted in Sect. 10.6 of [7] and the indicated additional references.

In the present case, a mask was created with the shape of the contour of the image (the mask boundary delimiting the region of interest is shown in Fig. 8.9 for the reference configuration in each coordinate plane). This mask made it possible to propagate the fringes outside the contour by an iterative process and finally produced results free from boundary errors. In the 3-D pattern analysis different masks must be created. These masks must be compatible with each other taking into consideration their mutual relationship in 3-D (for example, the thickness of the cardiac wall cannot change significantly passing from a coordinate plane to another). Since the ultimate end was to derive results that allow to evaluate for example volumes of the ventricle at different stages of the cycle care was taken in properly considering the anatomy of the walls at

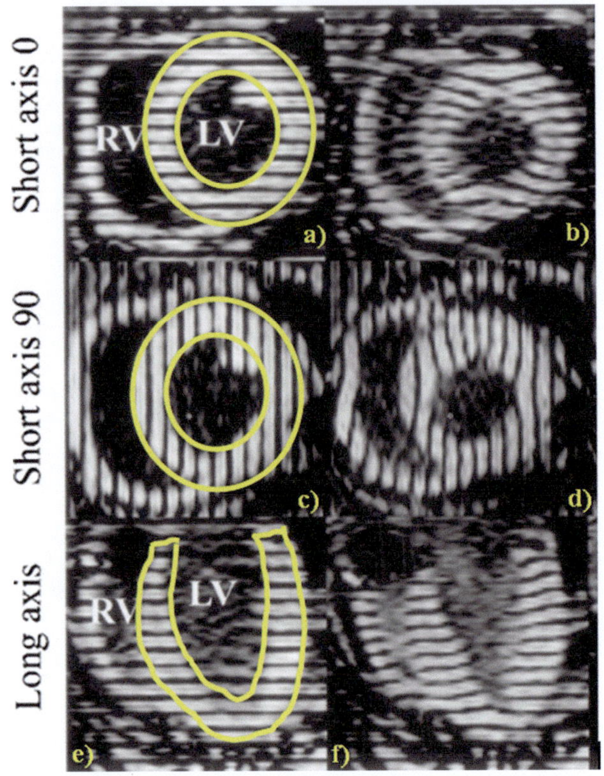

Fig. 8.9 Short-axis (0° and 90°) and long-axis MRI images of the right ventricle (RV) and left ventricle (LV). Reference configurations correspond to Fig. 8.9a, c, e. Images were taken from [11]

the different coordinate levels avoiding artifacts created by the different factors that influence the final images. Of course, selection of the masks outlines needs judgments in the choice of the profiles distorted by the different factors that influence the final images. The accuracy of the final results depends greatly on the quality of the particular MR system that is utilized. We will illustrate the main procedures by applying them to the images of Fig. 8.9. Later on we will discuss the possibility of a generalization and automation of the applied technique.

8.9 Processing of the Data Contained in MRI Tagged Images

The strains on the recording planes were computed utilizing the methodology explained in the previous sections. Since large deformations occur, the Eulerian approach to deformation analysis was applied in processing patterns of Fig. 8.9. For each recorded plane the principal and secondary strains were computed. The images of Fig. 8.9 correspond to the short axis of the left ventricle and to the long axis for the period of the contraction [11]. The reference images (Fig. 8.9a, c, e) were taken in the early systole while the analyzed image corresponds to the mid-systole. While rotations around the z axis are infinitesimal, the rotation of the long axis was corrected and scale correction between images were also done. Eulerian strains were computed using the simplified Almansi strain tensor. For the e_x^E, e_y^E and e_z^E components, it holds, respectively, [8]

$$e_x^E = \frac{\partial u}{\partial x} - \frac{1}{2}[(\frac{\partial u}{\partial x})^2 + (\frac{\partial v}{\partial x})^2 + (\frac{\partial w}{\partial x})^2] \tag{8.28}$$

$$e_y^E = \frac{\partial v}{\partial y} - \frac{1}{2}[(\frac{\partial u}{\partial y})^2 + (\frac{\partial v}{\partial y})^2 + (\frac{\partial w}{\partial y})^2] \tag{8.29}$$

$$e_z^E = \frac{\partial w}{\partial z} - \frac{1}{2}[(\frac{\partial u}{\partial z})^2 + (\frac{\partial v}{\partial z})^2 + (\frac{\partial w}{\partial z})^2] \tag{8.30}$$

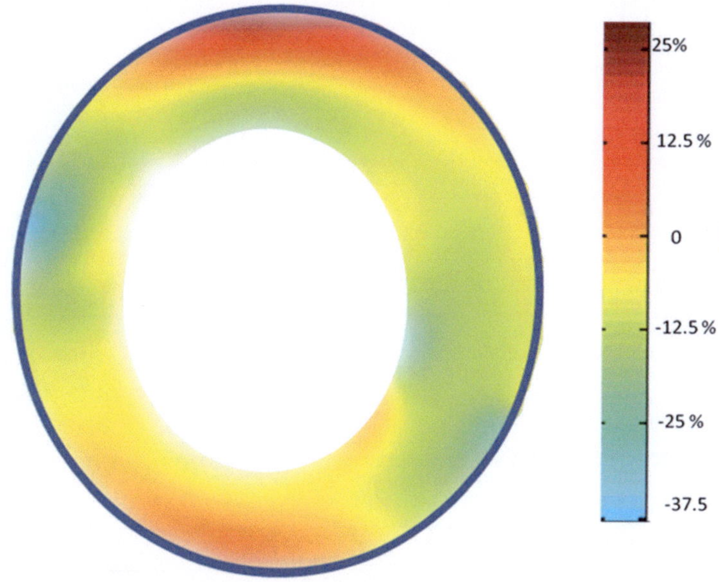

Fig. 8.10 Eulerian strains e_x^E of the left ventricle during the contraction period at mid-systole

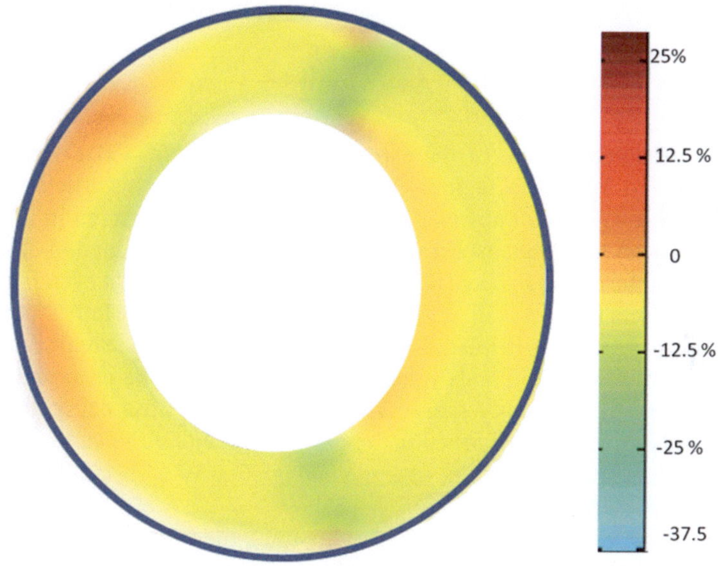

Fig. 8.11 Eulerian strains e_y^E of the left ventricle during the contraction period at mid-systole

Similar expressions were used for shear strains. For example, the generic strain component e_{ij}^E was computed as:

$$e_{ij}^E = \frac{1}{2}\left(\frac{\partial u_i}{\partial x_j} + \frac{\partial u_j}{\partial x_i}\right) - \frac{1}{2}\left[\frac{\partial u}{\partial x_i}\frac{\partial u}{\partial x_j} + \frac{\partial v}{\partial x_i}\frac{\partial v}{\partial x_j} + \frac{\partial w}{\partial x_i}\frac{\partial w}{\partial x_j}\right] \tag{8.31}$$

Figures 8.10 and 8.11, respectively, show the distributions of Eulerian strains e_x^E and e_y^E that develop along the short axis in the contraction period at mid-systole. Figure 8.12 shows the map of Eulerian strains e_z^E along the long axis at mid-systole. The contour of the mask limiting the region of interest corresponding to the left ventricle in each image pattern is marked in the figures.

Although in [11] no strain distributions were directly provided, time diagrams of circumferential strain are given. The Lagrangian strains were computed and the orders of magnitude of maximum strains in compression and tension agree well with the values indicated in Figs. 8.10, 8.11 and 8.12, maximum tension about 15 % maximum compression about 10 %. Although the Lagrangian strains have higher values than the Eulerian strains, the results obtained in this paper are in good agreement with the results provided in [11].

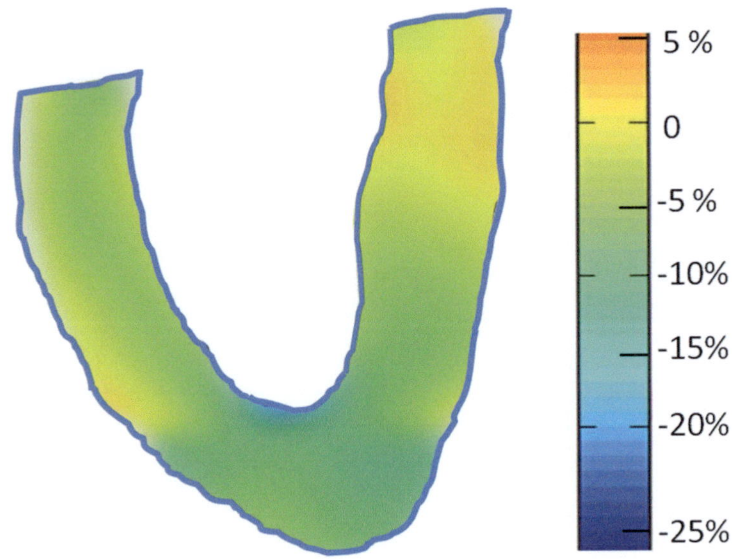

Fig. 8.12 Eulerian strains e_z^E of the left ventricle during the contraction period at mid-systole

8.10 Summary and Conclusions

The main principles of the experimental analysis of 3-D deformation on the basis of carrier planes that produce on the observed 3-D slices tagging lines were reviewed in this article. It is shown that the relative order determination makes it possible to connect any two images that have been captured in time without the need of any additional tracking procedure. This is an important aspect when one considers the automation of data analysis. It is also shown that with an adequate selection of the tagging planes it is possible to recover 3-D data even if one cannot get all the necessary families of tagging planes.

Effects that generate problems in the 3-D analysis volumes that have been tagged in general and in the particular in the case of MRI tagged images were discussed. Particular attention was given to rigid body rotations. A methodology for correction of 3-D rigid body rotations was developed. The proposed correction process simplifies computation on the basis of utilizing the frequency space rather than the actual physical space.

The processes described in the paper were applied to cardiac cycles images of the LV along the short axis and the long axis. The obtained results were in good qualitative agreement with the deformation data provided in the source paper of the analyzed images.

In conclusion it was shown that applying the techniques developed in the field of Experimental Mechanics it is possible to reduce the input of the human observer to a minimum in the analysis of tagged images. Of course it remains the task of concentrating in the MRI technique in order to get better quality tagged images reducing the task of merging sequentially obtained images and the volume of data that it is necessary to acquire to a minimum to characterize the mechanical aspects of the heart as a pumping device. In this way it will become feasible to use tagged images of the heart as a fast, reliable and economical diagnosis tool.

References

1. Sciammarella, C.A., Lamberti, L., Boccaccio, A.: Chapter 7: Data processing techniques to analyze large 3-D deformations of cardiac cycles. In: Jin, H., Sciammarella, C.A., Yoshida, S., Lamberti, L. (eds.) Advancement of optical methods in experimental mechanics, pp. 63–87. Springer, New York, NY (2015)
2. Sciammarella, C.A., Chiang, F.P.: The moiré method applied to three-dimensional elastic problems. Exp. Mech. **4**(11), 313–319 (1964)
3. Sciammarella, C.A., Di Chirico, G., Chang, T.Y.: Moiré holographic technique for three-dimensional analysis. J. Appl. Mech. **37**, 180–185 (1970)
4. Wright, G.A.: Magnetic resonance imaging. IEEE Signal Process. Mag. **14**(1), 56–66 (1997)
5. Lustig, M., Donoho, D.L., Santos, J.M., Pauly, J.M.: Compressed sensing MRI. IEEE Signal Process. Mag. **25**(2), 72–82 (2008)

6. Sciammarella, C.A., Lamberti, L.: Basic models supporting experimental mechanics of deformations, geometrical representations, connections among different techniques. Meccanica **50**(2), 367–387 (2015)
7. Sciammarella, C.A., Sciammarella, F.M.: Experimental mechanics of solids. Wiley, Chichester (2012)
8. Durelli, A.J., Parks, V.J.: Moiré analysis of strain. Prentice Hall, Englewood Cliffs, NJ (1971)
9. Ashburner, J., Friston, K.J.: Chapter 2: Rigid body registration. In: Frackowiak, R.S.J., Friston, K.J., Frith, C., Dolan, R., Price, C.J., Zeki, S., Ashburner, J., Penny, W.D. (eds.) Human brain function. Academic Press, San Diego, CA (2003)
10. Sciammarella, C.A., Lamberti, L., Sciammarella, F.M., Boccaccio, A.: The kinematics and dynamics of 3-D displacement fields. In: Jin, H., Sciammarella, C.A., Yoshida, S., Lamberti, L. (eds.) Advancement of optical methods in experimental mechanics, pp. 43–67. Springer, New York, NY (2014). Chapter 7
11. Wyman, B.T., Hunter, W.C., Prinzen, F.W., McVeigh, E.R.: Mapping propagation of mechanical activation in the paced heart with MRI tagging. Am. J. Physiol. **276**(3), H881–H891 (1999)

Chapter 9
A Deeper Look Into Immature Porcine Zona Pellucida Visco-hyperelasticity

Antonio Boccaccio, Luciano Lamberti, Massimiliano Papi, Marco De Spirito, and Carmine Pappalettere

Abstract The usual assumption made in mechanical characterization of soft biotissues with Atomic Force Microscopy (AFM) is that the specimen behaves as a purely elastic material. However, there is a limit indentation rate below which viscous effects can be neglected. A parametric study including about 200 FEM analyses shows that in the case of immature porcine zona pellucida (ZP) samples viscous effects become more significant for sharp tips. A linear relationship between the limit indentation rate and the geometry of the AFM probe is derived for the porcine ZP samples analyzed in this study.

Keywords Soft matter • Mechanical characterization • Atomic Force Microscopy • Probe geometry • Indentation rate • Viscous effects

9.1 Introduction

Atomic Force Microscopy (AFM) is an effective approach to mechanical characterization of soft biotissues such as cells and biopolymer networks [1–8]. Aspects concerned with calibration of nanoindentation measurements [9], sensitivity of finite indentation response to probe geometry [10] and residual stresses [11] also were investigated.

The typical assumption made in atomic force spectroscopy is that materials are purely elastic: different constitutive laws ranging from linear elasticity to hyperelasticity were hypothesized. Viscoelastic effects can be neglected if quasi-static tests are performed [12]. However, recent investigations [13] proved that viscous forces may dominate the nanoindentation response of biopolymer networks, such as zona pellucida (ZP) membranes, even at rather low indentation rate.

Experimental determination and modeling of viscoelastic behavior of soft materials is well documented in literature [13–22]. The present authors developed a visco-hyperelastic model to describe the nanoindentation response of immature porcine zona pellucida (ZP) membranes [23]. The hyperelastic behavior was described by the Arruda-Boyce constitutive law while the viscous response was defined by a N-terms Prony series expansion of the dimensionless relaxation modulus. By comparing finite element computations and AFM data at different indentation rates, it was possible to separate hyperelastic behavior and viscous effects.

This study will analyze in detail the sensitivity of the mechanical characterization process of immature porcine zona pellucida (ZP) membranes to nanoindentation conditions. In particular, we aim to determine any mathematical relationship between the probe geometry and the range of limit indentation rates for which it is possible not to account for visco-elastic effects.

9.2 Methodology

A parametric axisymmetric finite element model simulated the nanoindentation process. The general purpose finite element software ABAQUS® Version 6.12 (Dassault Systèmes, France) was utilized. The axis of symmetry of the FE model coincides with the direction of indentation. The indenter was modeled as a rigid blunt cone while the sample was modeled as an incompressible visco-hyperelastic slab with 60 μm diameter and 6 μm thickness (Fig. 9.1).

A. Boccaccio • L. Lamberti (✉) • C. Pappalettere
Dipartimento di Meccanica, Matematica e Management, Politecnico di Bari, 70126 Bari, Italy
e-mail: luciano.lamberti@poliba.it

M. Papi • M. De Spirito
Istituto di Fisica, Università Cattolica del Sacro Cuore, 00168 Rome, Italy

S.A. Tekalur et al. (eds.), *Mechanics of Biological Systems and Materials, Volume 6*,
Conference Proceedings of the Society for Experimental Mechanics Series, DOI 10.1007/978-3-319-21455-9_9

The bottom edge nodes of the model were clamped while the axis of symmetry could not displace in the horizontal direction: boundary conditions are indicated in Fig. 9.1. A vertical displacement of $\delta = 100$ nm was imposed to the rigid body part of the FE model simulating the AFM tip. The contact interaction between the AFM tip and the soft material sample was hypothesized to be frictionless and the hard contact option available in ABAQUS was selected.

Twelve different geometric configurations were considered in this parametric study. The AFM tip radius of curvature R was set as 10, 30 and 50 nm while the blunt cone angle of aperture α was set as 20°, 40°, 60° and 80°, respectively. Tip dimensions were consistent with commercially available products and accounted for changes of probe geometry that might occur in AFM measurements.

The sample was first modeled as a purely hyperelastic material following the Arruda-Boyce constitutive law [24], originally developed to simulate the mechanical response of polymeric chains. The same sample was then modeled as a visco-hyperelastic material where viscous behavior is described by a N-terms Prony series expansion of the dimensionless relaxation modulus. The effective relaxation modulus $\mu_{R8chain}(t)$ is the product between the instantaneous shear modulus μ_{8chain} and the dimensionless relaxation function.

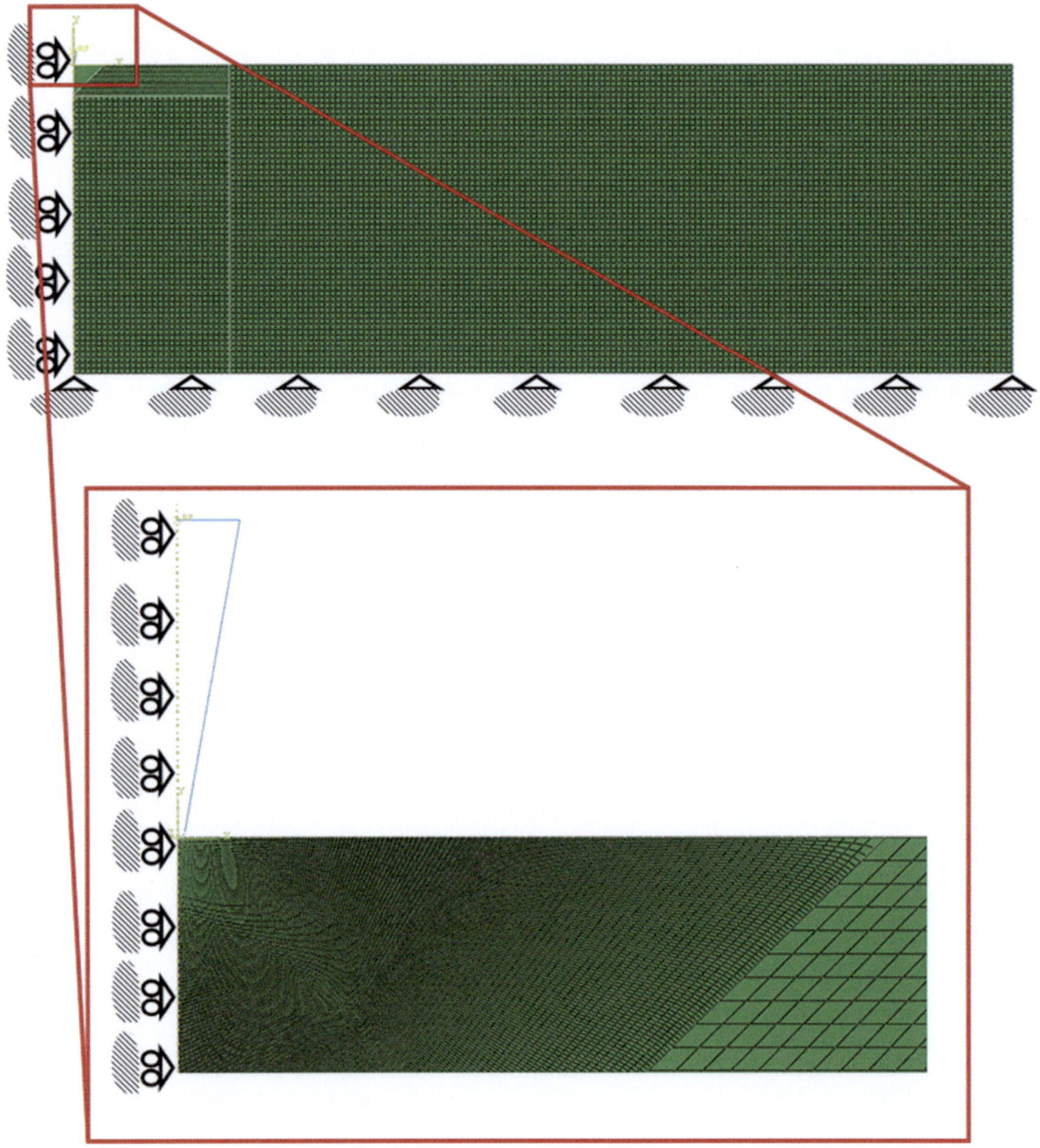

Fig. 9.1 Finite element model of the indented immature porcine ZP membrane samples

For the purely hyperelastic model and each AFM tip geometric configuration, one time-independent finite element analysis was performed by giving in input to ABAQUS the μ_{8chain} = 1.688 kPa, λL = 2.958 hyperelastic parameters determined in [23]. In the visco-hyperelastic model, for each AFM tip configuration, time-dependent FE analyses were performed for the indentation rates vi: 100, 200, 300, 400, 500, 1000, 2000, 3000, 4000, 5000, 6000, 7000, 8000, 9000 and 10,000 nm/s. The visco-hyperelastic parameters given in input to ABAQUS were μ_{8chain} = 1.688 kPa, λL = 2.958, $g1$ = 0.9681 and $\tau 1$ = 0.00133 s [23]. The parametric FE model was validated against experimental data gathered for some combinations of tip geometry and indentation rate.

The reaction forces $F_{\delta=100nm}^{visco-hyperelastic}$ and $F_{\delta=100nm}^{hyperelastic}$ (acting on the tip) computed by ABAQUS by assuming, respectively, the visco-hyperelastic behavior and the purely hyperelastic behavior were compared in order to find their difference

$$\varepsilon = \left| \frac{F_{\delta=100nm}^{visco-hyperelastic} - F_{\delta=100nm}^{hyperelastic}}{F_{\delta=100nm}^{hyperelastic}} \right| \times 100 \tag{9.1}$$

for each indentation rate and each AFM tip geometry considered in this study. The purely hyperelastic model was judged in agreement with the visco-hyperelastic model if $\varepsilon < 25$ %.

9.3 Results and Discussion

No systematic studies were ever carried out on the relationships between AFM tip geometry, indentation rate and visco-elastic response of immature porcine zona pellucida. In order to overcome this limitation, we developed a visco-hyperelastic model able to estimate the limit indentation rate $v_{threshold}$ below which viscous effects become negligible and a purely hyper-elastic model may reliably describe the material behavior.

A parametric axisymmetric finite element model simulated the contact between the indenter and the sample. The parameters considered in this study were the radius R of the AFM tip and the angle of aperture α of the blunt cone. For each geometric configuration of the AFM probe, one static hyperelastic analysis and 15 time-dependent visco-hyperelastic analyses (covering indentation rates ranging from 0.1 to 10 μm/s) were performed. The model was validated against experimental data gathered for some combinations of tip geometry and indentation rate.

It was found that the shape of the contact surface between the tested sample and the indenter strongly depends on the geometry of the AFM tip. In particular, deformations of indented samples were highly localized near the probe in the case of sharper tips. The percent difference ε between indentation forces computed by ABAQUS assuming a visco-hyperelastic behavior or a purely hyperelastic behavior increased considerably with indentation rate.

For all indentation rates, the percent difference ε between indentation forces becomes more significant for sharp tips. For a given value of the angle α, it increases as the tip radius R gets smaller. For a given value of tip radius, the percent difference ε drops down as the angle of aperture α of the blunt cone increases. Therefore, viscous effects appear to be more relevant in the case of sharp probes. The limit indentation rate $v_{threshold}$ changes accordingly, passing from about 180 nm/s evaluated for the very sharp tip (i.e., R = 10 nm and α = 20°) to about 430 nm/s for the very smooth tip (i.e., R = 50 nm and α = 80°).

The results obtained in this research can be explained as follows. Sharper indenters produce higher stress concentrations near the probe tip, in particular higher shear stresses. This triggers the viscous reaction of the polymeric networks that depends on the relative slippage of chains. While the "elastic" stiffness of biopolymers is driven by their network structure and on the ability of polymeric chains to deform independently one from another, the "viscous" part represents the equivalent stiffness of the spring-dashpot system describing the mechanical behavior of the material. The polymer reaction force developing at very low velocity is dominated by the elastic term. Viscosity increases the reaction force with velocity until the polymer reaction force reaches a plateau value.

Relationship between limit indentation rate and probe geometry for immature porcine ZP samples. An interesting question to be addressed is the following. Is it possible to establish for a given material a direct mathematical relationship between the limit indentation rate and the probe geometry? In order to answer this question, we averaged the limit indentation rates determined for the immature porcine ZP samples at each curvature radius and plotted them with respect to the radius of curvature of the AFM tip. Figure 9.2 shows the result of this operation. It can be seen that the average threshold velocity varies linearly with respect to the probe radius of curvature. Data are fitted by a linear equation at a very high degree of correlation (R2=0.9943). The etched curves shown in the plot correspond to the range of variation of the data gathered in the parametric study and also are very well fitted by linear equations. Preliminary tests carried out on ZP membranes of other species (in particular, immature equine ZPs that are considerably softer than immature porcine ZPs), seem

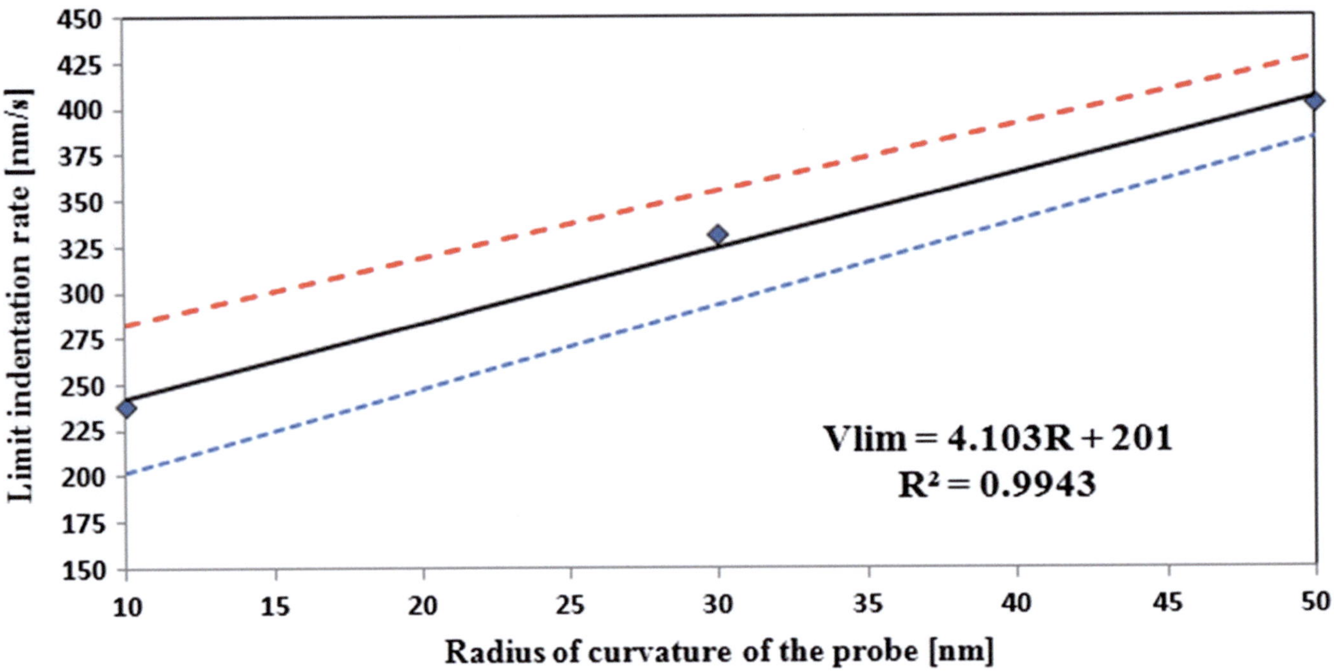

Fig. 9.2 Variation of the difference between visco-hyperelastic and hyperelastic model indentation forces for different AFM tip geometries and indentation rates

to confirm the trend observed in the present study although the range of limit indentation rates is obviously different due to the considerably different stiffness of the material to be characterized. Generally speaking, the above mentioned linear relationship may be a direct consequence of having assumed a linearly visco-hyperelastic constitutive behavior. However, more extensive investigations have to be carried out in order to address this important question.

9.4 Conclusions

This study analyzed with a great deal of detail the effect of viscous forces on the mechanical characterization of immature porcine zona pellucida samples with atomic force spectroscopy. A parametric finite element study proved that for each tip geometry there is a limit indentation rate $v_{threshold}$ beyond which the effect of viscosity cannot be neglected. Since viscosity effects appear to be more significant for sharp tips, using smooth profile indenters and reasonably low indentation rates will minimize errors on mechanical properties.

The average threshold velocity was found to vary linearly with respect to the radius of curvature of the AFM tip. Preliminary tests carried out on ZP membranes of other species (in particular, immature equine ZPs that are considerably softer than immature porcine ZPs) seem to confirm the trend observed in the present study although the range of limit indentation rates is obviously different from that found in this research (180-430 nm/s) due to the considerably different stiffness of the material to be characterized.Remarkably, the novel approach to sensitivity analysis of mechanical characterization of soft biotissues described in the article is very general and can be applied in principle to any material.

References

1. Suresh, S.: Biomechanics and biophysics of cancer cells. Acta Biomater. **3**(4), 413–438 (2007)
2. Tang, Y., Chen, X., Yoo, J., Yethiraj, A., Cui, Q.: Numerical simulation of nanoindentation and patch clamp experiments on mechanosensitive channels of large conductance in Escherichia coli. Exp. Mech. **49**(1), 35–46 (2009)
3. Ikai, A.: A review on: atomic force microscopy applied to nano-mechanics of the cell. Adv. Biochem. Eng. Biotechnol. **119**, 47–61 (2010)

4. Christ, A.F., Franze, K., Gautier, H., Moshayedi, P., Fawcett, J., Franklin, R.J.M., Karadottir, R.T., Guck, J.: Mechanical difference between white and gray matter in the rat cerebellum measured by scanning force microscopy. J. Biomech. **43**(15), 2986–2992 (2010)
5. Jacot, J.G., Martin, J., Hunt, D.L.: Mechanobiology of cardiomyocyte development. J. Biomech. **43**(1), 93–98 (2010)
6. Bernick, K.B., Prevost, T.P., Suresh, S., Socrate, S.: Biomechanics of single cortical neurons. Acta Biomater. **7**(3), 1210–1219 (2011)
7. Papi, M., Brunelli, R., Familiari, G., Frassanito, M.C., Lamberti, L., Maulucci, G., Monaci, M., Pappalettere, C., Parasassi, T., Relucenti, M., Sylla, L., Ursini, F., De Spirito, M.: Whole-depth change in bovine zona pellucida biomechanics after fertilization: how relevant in hindering polyspermy? PLoS One **7**(e45696), 1–7 (2012)
8. Boccaccio, A., Frassanito, M.C., Lamberti, L., Brunelli, R., Maulucci, G., Monaci, M., Papi, M., Pappalettere, C., Parasassi, T., Sylla, L., Ursini, F., De Spirito, M.: Nanoscale characterization of the biomechanical hardening of bovine zona pellucida. J.R. Soc. Interface **9**(76), 2871–2882 (2012)
9. Barone, A.C., Salerno, M., Patra, N., Gastaldi, D., Bertarelli, E., Carnelli, D., Vena, P.: Calibration issues for nanoindentation experiments: direct atomic force microscopy measurements and indirect methods. Microsc. Res. Tech. **73**(10), 996–1004 (2010)
10. Costa, K.D., Yin, F.C.P.: Analysis of indentation: implications for measuring mechanical properties with atomic force microscopy. J. Biomech. Eng. **121**(5), 462–471 (1999)
11. Boccaccio, A., Papi, M., De Spirito, M., Lamberti, L., Pappalettere, C. Effect of the residual stress on soft sample nanoindentation. Appl. Phys. Lett. 102, 133704, 1–4 (2013)
12. Radmacher, M.: Measuring the elastic properties of living cells by the atomic force microscope. In: Jena, B.J., Horber, J.K.H. (eds.) Atomic Force Microscopy in Cell Biology, vol. 68, pp. 67–90. Elsevier, San Diego, CA (2002)
13. Papi, M., Maiorana, A., Douet, C., Maulucci, G., Parasassi, T., Brunelli, R., Goudet, G., De Spirito, M. Viscous forces are predominant in the zona pellucida mechanical resistance. Appl. Phys. Lett. 102, 043703, 1–4 (2013)
14. Guilak, F., Tedrow, J.R., Burgkart, R.: Viscoelastic properties of the cell nucleus. Biochem. Biophys. Res. Commun. **269**, 781–786 (2000)
15. Mathur, A.B., Collinsworth, A.M., Reichert, W.M., Kraus, W.E., Truskey, G.A.: Endothelial, cardiac muscle and skeletal muscle exhibit different viscous and elastic properties as determined by atomic force microscopy. J. Biomech. **34**(12), 1545–1553 (2001)
16. Trepat, X., Grabulosa, M., Puig, F., Maksym, G.N., Navajas, D., Farré, R.: Viscoelasticity of human alveolar epithelial cells subjected to stretch. Am. J. Physiol. Lung Cell. Mol. Physiol. **287**(5), L1025–L1034 (2004)
17. Puig-de-Morales-Marinkovic, M., Turner, K.T., Butler, J.P., Fredberg, J.J., Suresh, S.: Viscoelasticity of the human red blood cell. Am. J. Physiol. Cell Physiol. **293**(2), C597–C605 (2007)
18. Darling, E.M., Topel, M., Zauscher, S., Vail, T.P., Guilak, F.: Viscoelastic properties of human mesenchymally-derived stem cells and primary osteoblasts, chondrocytes, and adipocytes. J. Biomech. **41**(2), 454–464 (2008)
19. Lubarda, V.A., Marzani, A.: Viscoelastic response of thin membranes with application to red blood cells. Acta Mech. **202**(1–4), 1–16 (2009)
20. Lee, Y.J., Patel, D., Park, S.: Local rheology of human neutrophils investigated using atomic force microscopy. Int. J. Biol. Sci. **7**(1), 102–111 (2011)
21. Pravincumar, P., Bader, D.L., Knight, M.M. Viscoelastic cell mechanics and actin remodelling are dependent on the rate of applied pressure. PLoS One 7, e43938, 1–10 (2012)
22. Jérusalem, A., Dao, M.: Continuum modeling of a neuronal cell under blast loading. Acta Biomater. **8**(9), 3360–3371 (2012)
23. Boccaccio, A., Lamberti, L., Papi, M., De Spirito, M., Douet, C., Goudet, G., Pappalettere, C. A hybrid characterization framework to determine the visco-hyperelastic properties of a porcine zona pellucida. Interface Focus 4(2), 20130066, 1–10 (2014)
24. Arruda, E.M., Boyce, M.C.: A three-dimensional constitutive model for the large stretch behavior of rubber elastic materials. J. Mech. Phys. Solids **41**(2), 389–412 (1993)

Chapter 10
Induced Coupling Between Contractile Cells

Brian J. Williams and M. Taher A. Saif

Abstract Cells communicate with each other through biochemical and electrical mechanisms employing various autocrine and endocrine signaling pathways. Cells, and cardiomyocytes in particular, generate contractile forces on the substrates (or tissues) they adhere to. This results in a tensile strain field in the substrate around the contractile cell. As a result, nearby cells adhered to the same substrate get stretched. The effect of this long range communication has received limited attention to date. Here we develop an elastic cell culture substrate to explore strain mediated coupling in cardiomyocytes. We investigate (1) whether strain coupling of isolated cardiomyocytes affects the temporal dynamics of contractility, (2) whether such long range communication can enable synchrony in cardiomyocyte beating, and (3) what is the biophysical mechanism enabling strain induced coupling? Initial experiments show that strain coupled populations of neonatal rat cardiomyocytes can synchronize their beating with time. We hypothesize that the contraction of one cell cluster results in calcium influx in the coupled cell cluster via stretch sensitive ion channels. We utilize an integrate-and-fire oscillator model with strain induced coupling to predict the emergence of synchrony between cardiomyocytes. This finding may shed light on cardiac arrhythmias in stiffened, infarcted cardiac tissues where strain coupling may be compromised.

Keywords Synchronization • Cardiomyocytes • Strain-coupling • PIV • Oscillators

10.1 Introduction

In a functioning heart, millions of cardiomyocytes, the contractile cells in cardiac tissue, must beat in synchrony to pump blood through the circulatory system. The dominant mechanism enabling this synchronized contraction is electrical coupling through gap junctions, where the depolarization of nodal pacemaker cells is transmitted to "working" cells by specialized conducting cells. Many studies have established the role of stretch in calcium handling in working myocytes [1–6], but, due to experimental limitations, the effect of cyclic stretch on the temporal contractile dynamics of cardiomyocytes is not well understood. Mechanical strain travels through cardiac tissue at over 1600 m/s [7], compared to membrane depolarization at 0.5–2 m/s by electrical coupling [8]. Here, we propose that the mechanical stretch of cardiomyocytes induced by the contraction of other myocytes in the same organ may provide a supplementary synchronizing mechanism in the contractile dynamics of myocardial tissue (Fig. 10.1). Such a supplementary mechanism may improve the level of synchronization or provide coupling in cases of limited gap junction connectivity.

Studying extracted cardiomyocytes in vitro bypasses the normal pacemaker to conducting cell to working cell chain. Each cardiomyocyte can function as an independent oscillator [9]. At the beginning of the cycle, the myocyte maintains a membrane potential of −85 to −95 mV while generating no contractile force. The cell undergoes a slow diastolic depolarization via calcium influx ($I_{Ca^{2+}}$) and sodium influx ($I_{Na^{+}}$) until a full action potential is triggered. The cell rapidly depolarizes with Ca^{2+} and Na^{+} influx. Increased intracellular calcium concentration ($[Ca^{2+}]i$) triggers the release of Ca^{2+} from the sarcoplasmic reticulum (SR). Myosin motors in the sarcomeric structures within the cell, activated by Ca^{2+}, apply contractile force on the z-disks by holding onto the actin fiber bundles. The z-disks approach each other, and the cell contracts (systolic phase). $[Ca^{2+}]i$ is depleted by transport out of the cell or into the SR and mitochondria. Myosin motors release the actin, and the cell relaxes (diastolic phase).

B.J. Williams • M.T.A. Saif (✉)
University of Illinois at Urbana-Champaign, 1206 W Green St, Urbana, IL 61801, USA
e-mail: saif@illinois.edu

S.A. Tekalur et al. (eds.), *Mechanics of Biological Systems and Materials, Volume 6*,
Conference Proceedings of the Society for Experimental Mechanics Series, DOI 10.1007/978-3-319-21455-9_10

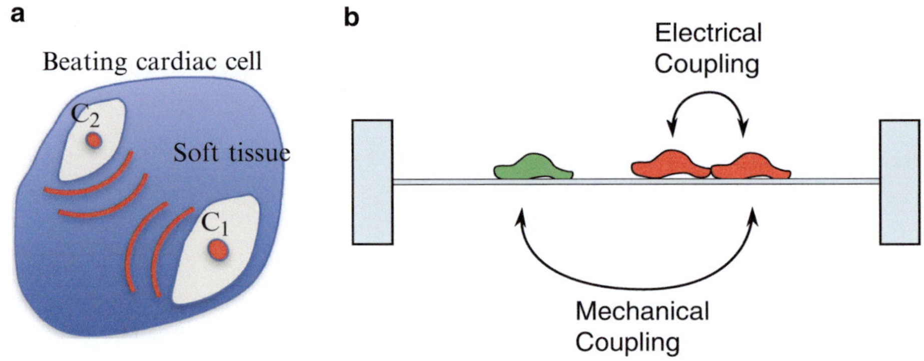

Fig. 10.1 (**a**) Contractile cells on soft tissues induce deformation in the substrate that stretch or compress nearby cells, and (**b**) we seek to study the effect of periodic mechanical strain on the temporal dynamics of spontaneously contracting cells in the absence of electrical coupling via gap junctions

A cluster of spontaneously contracting cardiomyocytes are triggered by whichever cells fire first via diastolic depolarization. Calcium dynamics in cardiomyocytes show strong stretch sensitivity. Intracellular calcium ($[Ca^{2+}]i$) increases linearly with stretch [1]. This increase can be significantly inhibited by blocking L-type Ca^{2+} channel and stretch activated channels using diltiazem and gadolinium, respectively. Increase in $[Ca^{2+}]i$ above a threshold leads to more Ca^{2+} release from the SR, activation of sarcomeric myosin, and cell contraction. As a result, stretching a cluster of spontaneously contracting cardiomyocytes may induce increased Ca^{2+} influx and a more rapid diastolic depolarization, thus increasing the firing rate of the triggering cell and all other cells electrically coupled to it. We propose that, if multiple electrically isolated cardiomyocyte clusters induce stretch onto each other by deforming the substrate between them, then they should influence each others dynamics, i.e., they should become coupled. The strength of this coupling depends on the sensitivity to stretch and the amount of stretch itself. The potential to synchronize depends on the coupling strength, the variation between uncoupled contractile rates of individual clusters, and stochastic noise in the system.

We have previously demonstrated that mechanical stimulation can induce contractility in quiescent cardiomyocytes. Primary chicken cardiomyocytes, extracted from embryonic chicken hearts, were cultured on inert soft 2D polyacrylamide (PA) substrates [10]. During culture, some of the cardiac myocytes beat, while others occasionally beat or keep quiescent for most of the culture time. A rigid tungsten probe was mounted on a piezo stage with xyz actuators (Fig. 10.2a). The probe contacted the gel surface about 60 μm away from the cell and exerted a small compressive force on the gel. The probe was then used to cyclically stretch the substrate (Fig. 10.2b) in a pulsing fashion with varied frequency, from 20 to 45 cycles per minute, causing the cell to stretch by about 6 %. Thus, mechanical stretch was applied to the cell through the substrate only. After seven cycles of such stretches (about 20 s), the quiescent cell started to beat, and continued to beat for several hours after the mechanical stimulation was stopped. Out of the fifteen quiescent cells that were stimulated by the probe, 8 were found to beat. The average stretch applied by the probe on the cells was 5.89 ± 0.62 % of their original size.

These results indicate that chicken cardiomyocytes are sensitive to mechanical stimulation. This experiment suggests that if the probe were replaced by a beating cardiomyocyte, then the beating cell may also stimulate a neighbor by stretching and releasing the soft substrate. Such stimulation may 'awaken' the quiescent neighbor, which in turn may stimulate the neighbor that awoke it in the first place. Thus the beating of the pair may become coupled, and they may continue to beat for a longer time than isolated cells.

10.2 Experiment

We developed a simple test platform to test for synchrony between cardiac cells through substrate elasticity. The platform is a simple PDMS membrane (Fig. 10.3a, b), 10–20 μm thick, supported by a glass ring. 10:1 PDMS (Sylgard 184, Dow Corning) is spun on a polystyrene substrate at 6000 RPM for 30 s and cured at 60 °C for 6 h. A 17 mm hole is drilled into a 23 × 23 mm, 1 mm thick glass slide. The glass slide is glued to the cured PDMS film with a thin bead of 10:1 PDMS and cured at 60 °C. The film is cut around the glass slide, and the slide is gently peeled from the substrate, producing a thin, highly compliant substrate with fixed boundary conditions. A second drilled glass slide is glued to the opposite side of the film to provide a "seeding well" on both sides. This configuration ensures long range strain coupling. The film compliancy can be

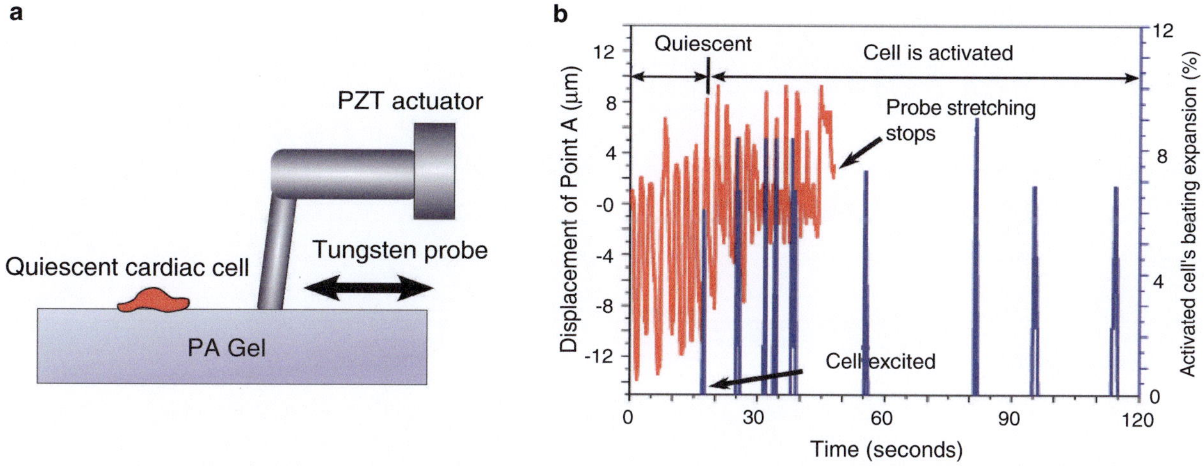

Fig. 10.2 Contractile activity can be induced in quiescent cardiac myocytes by stretching them. (**a**) Apparatus to mechanically stimulate cardiomyocytes, consisting of cells cultured on a soft PA gel substrate, and a tungsten probe attached to a micromanipulator to apply local substrate strain, and (**b**) periodic substrate strain induces contractile activity in cells that persists after mechanical stimulation is removed

tuned by varying the cross-linking ratio of the PDMS or the film thickness via modulation of spin-coating parameters. Films are functionalized by O_2 plasma exposure followed by incubation in fibronectin (25 μg/mL, Corning) for 30 min at RT.

Neonatal cardiomyocytes are extracted by established protocols [11]. Briefly, intact hearts are excised from 2 to 5 day old Sprague Dawley rat pups, minced into approximately 1 mm cubes, and digested in 0.05 % trypsin (Worthington Biochemical) at 4 °C for 16 h. The tissue is rinsed in HBSS, and digested in 0.1 % collagenase (Worthington Biochemical) at 36 °C for 45 min. The cells are mechanically dissociated and plated at approximately 250,000 cells/cm^2. Cells are cultured in high glucose DMEM, 2 mM L-glutamine, 10 mM HEPES, 10 % fetal bovine serum, and 1 % penicillin-streptomycin. To inoculate the films, 500 μL of the cell suspension is pipetted on one side of the film (Fig. 10.3c, d), the film is stored at 37 °C for 3 h, flipped upside down onto 500 μL of fresh media, and 500 μL of cell suspension is pipetted on the opposite side of the film (Fig. 10.3e). Three hours later, the entire slide is inundated in culture media. Media is changed every 48 h. A small glass spacer is used to tilt the film, permitting both sides of the film to access fresh media. Top and bottom cell suspensions can be stained with CellTracker Red and Green (Life Technologies) to distinguish between sides (Fig. 10.3f).

Within 2 days after plating, cells on either side of the PDMS film synchronize to each other by conventional (electrical) means. Effectively, each side of the film is acting as a single oscillator. The collective contraction generates large in-plane displacements of the film (Fig. 10.3g). To study the interaction of temporal dynamics of each oscillator, we need to extract contractile waveforms from both the top and bottom of the film. Videos are recorded of small, approximately 1 mm patches of the film at 50 frames per sec. A particle image velocimetry algorithm [12] is used to generate velocity fields for the film (Fig. 10.4a). To remove rigid body motion, a reference waveform is extracted from a region of the film with no contractile cells (Fig. 10.4b). Several waveforms are measured from the top and bottom of the film (Fig. 10.4c) by integrating velocity fields from representative contractile cells. Displacement amplitudes vary from cell to cell; all waveforms are normalized in amplitude for study of temporal dynamics. Representative waveforms from a single side of the film have closely matched temporal dynamics, indicating strong synchronization, while substantial variance is observed when comparing waveforms from the top and bottom of the film.

Due to variation in the contractile shapes from cycle to cycle, contractile waveforms must be further postprocessed to analyze oscillator interaction. We convert to phase space by choosing the point of maximal displacement to have a phase value of zero. We identify the time of maximal displacement by peak detection (Fig. 10.5a). Since we have no other tools to measure phase propagation between peaks, we assume linear phase propagation between zero-values (Fig. 10.5b). Finally, we use the phase difference between the top and bottom of the film to study phase interaction (Fig. 10.5c). A constant, zero phase difference indicates synchronization with zero phase lag, and a constant, nonzero phase difference indicates a phase shifted synchronization. A pair of constant but different frequency oscillators would exhibit a sawtooth relative phase.

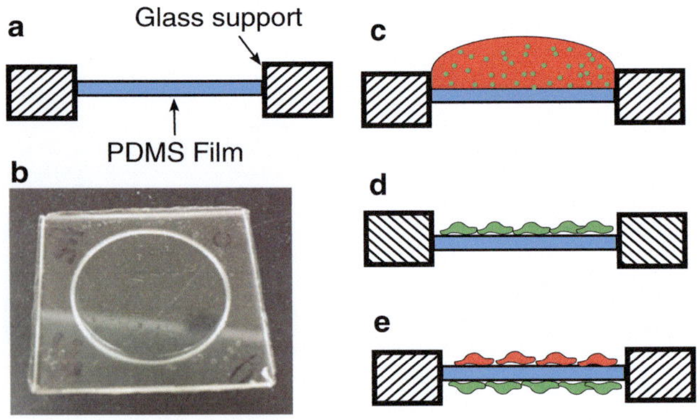

Fig. 10.3 PDMS film design. (**a**, **b**) Compliant substrates are fabricated by spin-coating a thin, 10–20 μm thick film of uncured PDMS on a polystyrene substrate, glueing a glass ring onto the film, curing the film, and peeling the film off the substrate. This produces a freestanding film under natural tension with fixed boundary conditions, (**c**–**e**) cells are plated on both sides of the film by inoculating the top surface, allowing cells to adhere for 3 h, flipping the substrate, and inoculating the opposite surface, (**f**) cells on either side of the film can be identified by staining with CellTracker *Red* or *Green*, and (**g**) cells on each side synchronize to each other by electrical means, producing in plane displacement of the film and inducing strain fields on cells on the opposite side of the film

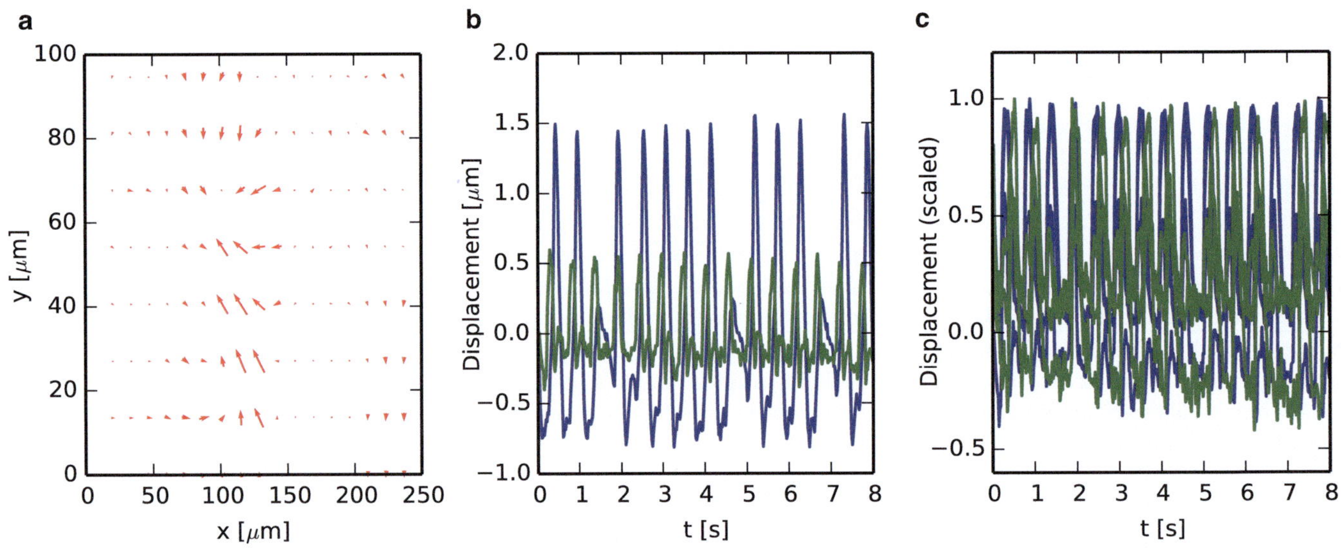

Fig. 10.4 Extraction of temporal dynamics from video. (**a**) A particle image velocimetry (PIV) algorithm is used to get the velocity field on the film. These velocity fields are integrated to generate displacement fields. (**b**) Contractile waveforms are calculated by taking the rigid body motion of the film (*blue*) from the tracked displacement function of a contractile cell, producing the local contractile displacement function (*green*). (**c**) Several displacement functions are taken from the top (*green*) and bottom (*blue*) of the film, demonstrating that cells on either side are in sync with each other, but not necessarily with cells on the opposite side of the film

10.3 Results

Two coupled, chaotic oscillators will synchronize if their coupling strength is sufficient to overcome intrinsic variation in natural oscillator frequency and random noise in oscillator dynamics. For synchronization to occur between our populations of cardiomyocytes, the proposed coupling mechanism must exist, and must be strong enough to overcome variation between uncoupled contractile rates. Not surprisingly, we observe significant variation in the interaction dynamics from experiment to experiment. Figure 10.5c plots the relative phase from a pair of oscillators (Fig. 10.5a) that synchronize. The green waveform tends to trail the blue waveform by a quarter cycle. The lagging oscillator occasionally misses a beat, only to fall back into synchronization on the next cycle in a phenomena observed in loosely coupled oscillators called phase slip [13].

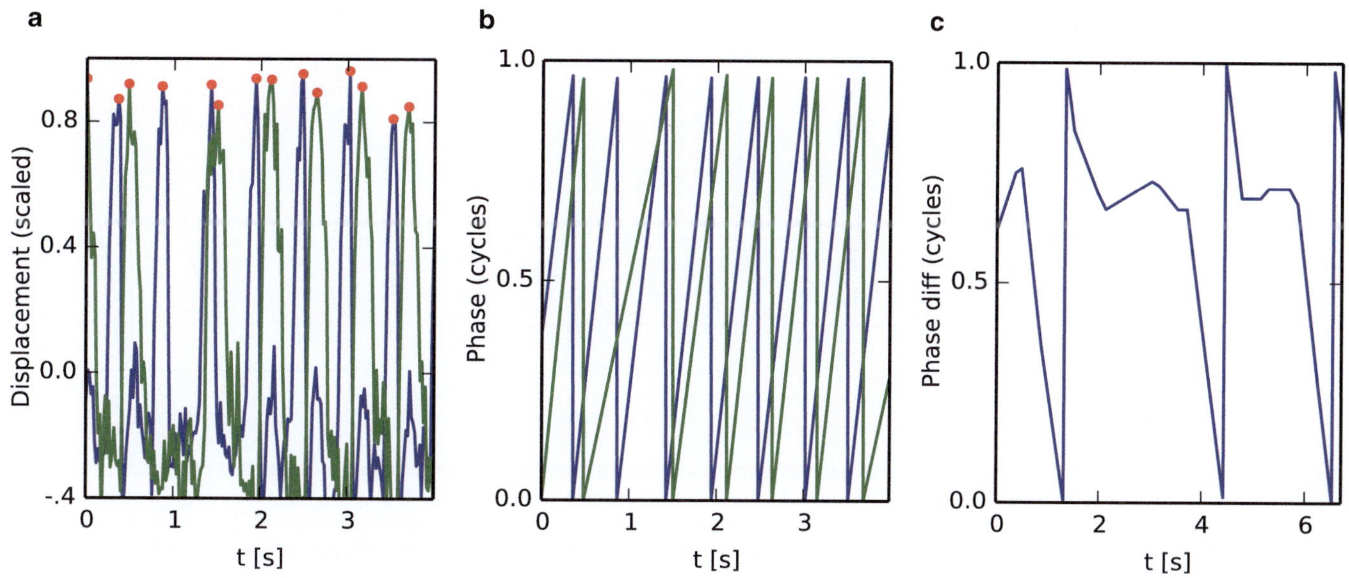

Fig. 10.5 Conversion to phase. (**a**) Phase values of 1 are assigned to the temporal points of maximal displacement in a given contractile cycle via peak detection of the associated displacement function (*red dots*), (**b**) representation of the temporal dynamics of the oscillators in phase space, where the phase is assumed to linearly increase between contractions, and (**c**) Analysis of oscillator interaction via study of the relative phase plot. Note, this dataset illustrates a sample that is in synchronization with occasional phase slips, as the *green* oscillator lags the *blue* oscillator by a quarter cycle with occasional missed contractions

In other experiments, the oscillator pair did not phase lock, and the relative phase propagates monotonically. Even in the absence of synchronization, phase interaction can be studied in the form of relative phase propagation velocity dependency on relative phase. Figure 10.6a plots one example where the relative phase tends to stutter at approximately 0.8. This reduction in slope indicates a slowing of relative phase propagation, potentially indicating that, for this relative phase, the leading oscillator is applying a "pull" on the trailing oscillator, before it completely overtakes it and the relative phase cycle repeats. Note, each cycle of the relative phase consists of several dozen contractile cycles (see Fig. 10.5).

To better understand how relative phase dependent coupling can effect relative phase propagation, we model the oscillator pair as two loosely coupled integrate-and-fire chaotic oscillators. Each oscillator accumulates Ca^{2+} until it exceeds a threshold, fires, and depletes its intracellular Ca^{2+} (Fig. 10.6b). Each model oscillator has steady and random Ca^{2+} influx and an efflux rate proportional to $[Ca^{2+}]i$. Each oscillator accounts for Ca^{2+} entry in cardiomyocytes due to stretch activated ion channels. When one cell fires, the other cell absorbs additional Ca^{2+}, with an amount that is dependent on its current phase. This phase windowing and coupling strength are manipulated as model parameters. The equation governing for the Ca^{2+} dynamics is as follows:

$$\dot{c}(x,t) = \dot{c}_s + \dot{c}_r + k_m(c)\varepsilon(x,t) - k_l c(x,t) \tag{10.1}$$

where $c(t)$ is the scaled Ca^{2+} concentration at time t, $\dot{c}_s$ is the steady influx, $\dot{c}_r$ is random influx to keep the system noisy, $\varepsilon(t)$ is the strain of the basal surface of the cell (and hence the substrate), km is the stretch sensitivity factor, and kl is the leakage factor, approximating Na/Ca exchangers removing intracellular Ca^{2+} at a rate proportional to $[Ca^{2+}]i$. Above the threshold concentration, the cell undergoes a contractile cycle and resets c to zero. The strain ε needs to be solved from an independent condition depending on the definition of the problem. For example, ε can be produced by the cells themselves when they are on an elastic substrate. In that case, ε is coupled with the calcium dynamics. When Ca^{2+} exceeds the threshold value, the cell generates a dipole, Pd, on the substrate, the value of which is available in the literature or can be measured by traction force microscopy. The strain field can then be evaluated by solving the elasticity problem involving the substrate and the dipole. In a system with multiple cells, the strain field due to all the cells can be computed by marching along time, and their beating pattern can be established. Substrate strain can also be prescribed if it is externally applied. In that case, $\varepsilon(t)$ is known and can be used to solve for $c(t)$ from Eq. (10.1).

Figure 10.6c illustrates the relative phase propagation for a pair of oscillators where the strain coupling is windowed by a narrow gaussian function centered at a phase lag of 0.2. In other words, if one oscillator trails the other by 0.2 cycles, it is

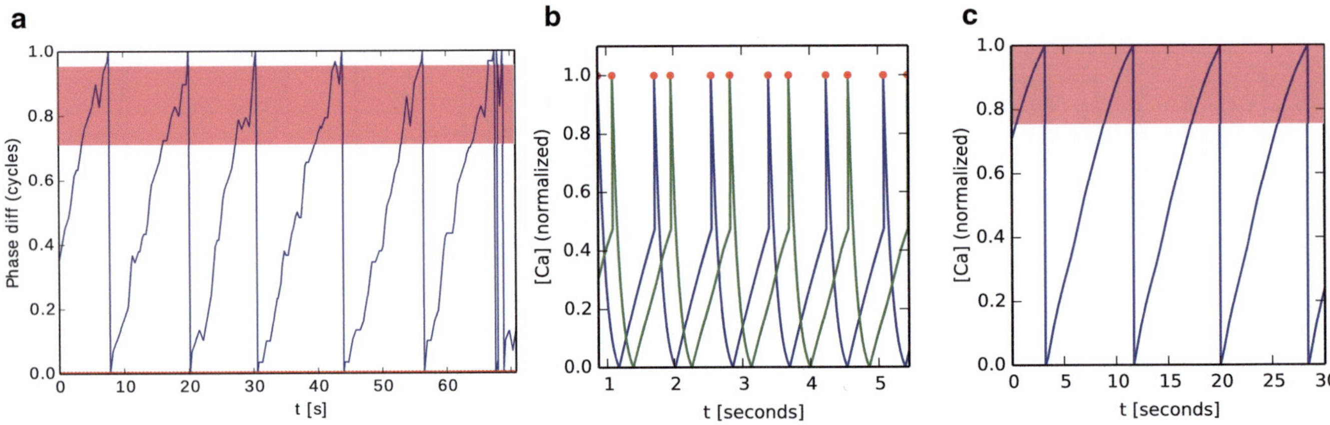

Fig. 10.6 Analysis of oscillator interaction in the absence of synchronization. (**a**) In one case, the relative phase function of an unsynchronized oscillator showed a tendency to stutter when the phase lag had a value of approximately of 0.8. Suggestive of relative phase dependent interaction, we model the system in (**b**), with a pair of integrate and fire oscillators, with slightly different natural frequencies, coupled by a Gaussian windowed function when one cell fires. This produces similar deviations in the *pink region* of (**c**), where relative phase propagation slows as the phase difference approaches 1.0

more susceptible to strain based coupling. The result is a flattening of relative phase propagation at relative phase values between 0.7 and 1, similar to what is observed experimentally. By modeling and analyzing cardiomyocytes as abstract oscillators in phase space, we gain access to a rich set of tools to study the dynamics of the constituent oscillators. Consider Fig. 10.4c, where the relative phase tends to be constant, with a value of approximately 0.7, for several contractile cycles at a time, until the trailing oscillator misses a fire, only to fall right back into step. This phase slip [14] can be used to quantify coupling strength as a function of phase slip frequency and recovery rate. In an unrelated physical system, this analysis method was used to quantify the coupling strength between a pair of beating flagella [13]. We plan to use the same techniques to quantify coupling between cardiac oscillators via specific ion channels and other myocyte components to determine the critical components enabling the observed coupling.

10.4 Conclusion

We have observed synchronization with phase slips in a pair of coupled cardiac clusters. With additional data, analysis of the dynamics of phase slips can be used to quantify the strength of oscillator interaction [13]. Further experiments will use molecular tools to attempt to identify the specific biological mechanisms enabling synchronization. Stretch activated channels can be suppressed with gadolinium [1], L-type calcium channels can be blocked with Diltiazem [1] or gallopamil [6], and sarcoplasmic reticulum ryanodine receptors can be blocked with Ruthenium Red and procaine [1].

Other data indicates that the relative phase of an oscillator pair can influence relative phase propagation. This can be a very useful tool in isolating the responsible biological mechanisms facilitating strain coupling. Different ion channels and other calcium handling mechanisms have well documented variations in relative significance at different phases of diastolic depolarization in the unstimulated cell [9]. By using relative phase dependency, we may be able to identify likely strain sensitive pathways based on which are dominant depolarizing agents in the associated phase of diastolic depolarization. Furthermore, variation in relative phase propagation is a quantifiable metric that may permit more detailed calculation of coupling strength in the absence of phase slips and full synchronization.

In summary, we have developed a platform to study strain coupling of physically isolated clusters of cardiac myocytes, provided a method to study phase coupling, identified synchronization in mechanically coupled cardiomyocytes, and identified potential relative phase dependence of relative phase propagation. Further study may provide results of biological and medical significance, as medical modulation of strain sensitivity could be used to minimize arrhythmias caused by scar tissue formation in post myocardial infarction hearts where electrical conductivity is impaired.

Acknowledgements This project was funded by the National Science Foundation (NSF), Science and Technology Center on Emergent Behaviors in Integrated Cellular Systems (EBICS) Grant CBET-0939511.

References

1. Ruwhof, C., Van Wamel, J., Noordzij, L., Aydin, S., Harper, J., Van Der Laarse, A.: Mechanical stress stimulates phospholipase C activity and intracellular calcium ion levels in neonatal rat cardiomyocytes. Cell Calcium **29**, 73–83 (2001)
2. Kamkin, A., Kiseleva, I., Isenberg, G.: Stretch-activated currents in ventricular myocytes: amplitude and arrhythmogenic effects increase with hypertrophy. Cardiovas. Res. **48**, 409–420 (2000)
3. Gannier, F., White, E., Lacampagne, A., Garnier, D., Le Guennec, J.-Y.: Streptomycin reverses a large stretch induced increase in $[Ca^{2+}]i$ in isolated guinea pig ventricular myocytes. Cardiovas. Res. **28**, 1193–1198 (1994)
4. Gannier, F., White, E., Garnier, D., Le Guennec, J.-Y.: A possible mechanism for large stretch-induced increase in $[Ca2+]i$ in isolated guinea-pig ventricular myocytes. Cardiovas. Res. **32**, 158–167 (1996)
5. Sigurdson, W., Ruknudin, A., Sachs, F.: Calcium imaging of mechanically induced fluxes in tissue-cultured chick heart: role of stretch-activated ion channels. Am. J. Physiol. Heart Circ. Physiol. **262**, H1110–H1115 (1992)
6. Tatsukawa, Y., Kiyosue, T., Arita, M.: Mechanical stretch increases intracellular calcium concentration in cultured ventricular cells from neonatal rats. Heart Vessel. **12**, 128–135 (1997)
7. Saijo, Y., Tanaka, M., Okawai, H., Sasaki, H., Nitta, S.-I., Dunn, F.: Ultrasonic tissue characterization of infarcted myocardium by scanning acoustic microscopy. Ultrasound Med. Biol. **23**, 77–85 (1997)
8. Durrer, D, Van Dam, R.T., Freud, G., Janse, M., Meijler, F., Arzbaecher, R.: Total excitation of the isolated human heart. Circulation **41**, 899–912 (1970)
9. Brette, F., Leroy, J., Le Guennec, J.-Y., Sallé, L.: Ca^{2+} currents in cardiac myocytes: old story, new insights. Prog. Biophys. Mol. Biol. **91**, 1–82 (2006)
10. Tang, X., Bajaj, P., Bashir, R., Saif, T.A.: How far cardiac cells can see each other mechanically. Soft Matter **7**, 6151–6158 (2011)
11. Williams, B.J., Anand, S.V., Rajagopalan, J., Saif, M.T.A.: A self-propelled biohybrid swimmer at low Reynolds number. Nat. Commun. **5**, (2014)
12. Tseng, Q., Duchemin-Pelletier, E., Deshiere, A., Balland, M., Guillou, H., Filhol, O., Théry, M.: Spatial organization of the extracellular matrix regulates cell–cell junction positioning. Proc. Natl. Acad. Sci. **109**, 1506–1511 (2012)
13. Goldstein, R.E, Polin, M, Tuval, I.: Emergence of synchronized beating during the regrowth of eukaryotic flagella. Phys. Rev. Lett. **107**, 148103 (2011)
14. Pikovsky, A., Rosenblum, M., Kurths, J.: Synchronization: A Universal Concept in Nonlinear Sciences. Cambridge University Press, Cambridge (2003)

Chapter 11
In-Vivo Indentation Testing of Sheep Spinal Cord with Meninges

Haitao Zhang, Phillip Falkner, and Chad Cai

Abstract In spinal cord stimulation (SCS) therapy neuromodulation leads are interacting with human dura during and after their implantations. Knowledge of the mechanical properties of spinal cord with meninges is essential for lead design, development, and optimization. In this study, an in-vivo indentation test method was developed to assess the equivalent Young's Modulus of sheep spinal cord with meninges. Histological analysis of the spinal cord with meninges was conducted. In-vivo equivalent Young's Modulus was compared with published Young's Modulus of cadaver dura and spinal cord respectively. The equivalent modulus has a mean of 0.022 MPa and a standard deviation of 0.014 MPa. Compared to the elastic modulus of spinal cord and the elastic modulus of dura mater, the equivalent elastic modulus of spinal cord with meninges is much smaller.

Keywords In-vivo • Indentation • Spinal cord • Young's Modulus • Neuromodulator

11.1 Introduction

Neuromodulation is among one of the fastest-growing areas of the medical device industry. Neurostimulators are being used to provide therapies to patients with chronic pains, movement disorders, overactive bladders, and psychiatric disorder symptoms. In these therapies, well-controlled stimulation current is delivered to the target nerves through implantable leads. For an example, electrical stimulation therapy is used for spinal cord stimulation (SCS) (Fig. 11.1a) to manage chronic spinal pain, such as back pain and leg pain. An implantable electrical stimulation system delivers electrical pulses (or therapy) via a medical device, called lead, to target spinal cord nerves to manage pains [1–4]. A typical SCS system consists of a stimulator and leads (Fig. 11.1b).

Figure 11.2 shows a sheep spinal cord with meninges which consist of, from outmost to innermost, dura, arachnoid, and pia maters. Cerebrospinal fluid fills the space between the arachnoid mater and spinal cord under the dura. Human spinal cord and meninges have the same structure as sheep's. To deliver SCS therapy, a lead needs to be implanted in epidural space in spinal canal where the lead will interact with dura during and after implantation. To implant the lead to a targeted treatment area during SCS lead implant procedure, a physician need to adjust the lead distal end direction left and right by rotating the proximal end of lead while advance the lead forward. Knowledge of the mechanical properties of spinal cord with meninges is of great importance for lead design, development and optimization. Although the mechanical properties of each individual component have been characterized in the past [5–11], to the best of the authors' knowledge, the equivalent in-vivo mechanical properties of spinal cord with meninges have not been characterized.

In this study, an in-vivo indentation test method was developed to measure the force and displacement of indenter on sheep spinal cord with meninges. An equivalent in-vivo Young's modulus of spinal cord with meninges was then obtained.

H. Zhang (✉) • P. Falkner • C. Cai
Medtronic, 7000 Central Ave. NE, Minneapolis, MN 55432, USA
e-mail: haitao.zhang@medtronic.com

S.A. Tekalur et al. (eds.), *Mechanics of Biological Systems and Materials, Volume 6*,
Conference Proceedings of the Society for Experimental Mechanics Series, DOI 10.1007/978-3-319-21455-9_11

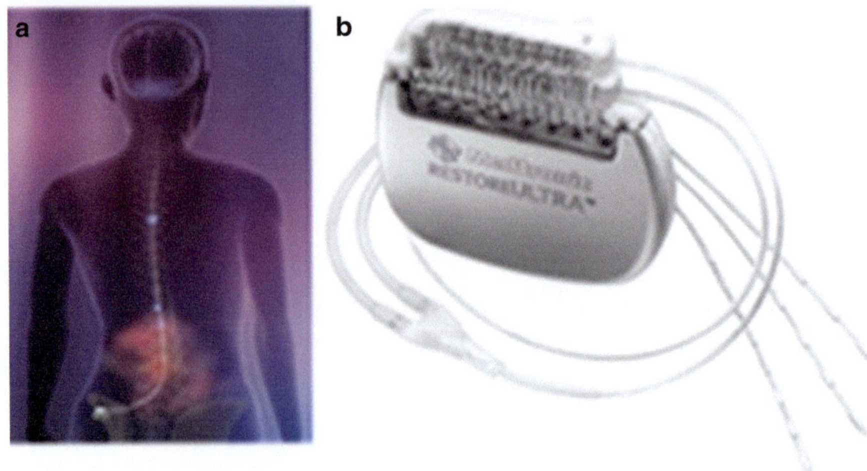

Fig. 11.1 Spinal cord stimulation and related system

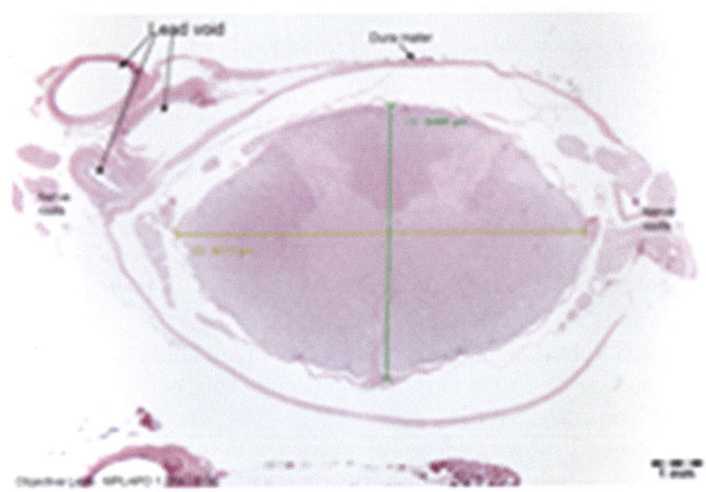

Fig. 11.2 Histological image of sheep spinal cord with meninges

11.2 Materials and Methods

11.2.1 Force and Displacement of Indenter on Dura

The force and displacement of the indenter on the spinal cord with meninges are measured by a customized testing system. This testing system was designed to measure displacement and force vs. time. It has a stepper motor connected to a Liner Electric Actuator (BIMBA OLE). At the end of the actuator there is a load cell to measure the force of resistance to the displacement. A 316 L stainless steel indenter is mounted to the load cell. The indenter is cylindrical with a spherical tip with radius of 2 mm, Fig. 11.3a, to minimize the adhesive effect of the dura surface, Fig. 11.3b. A Labview based software program was developed to control the stepper motor and measures displacement with the encoder mounted on the motor's shaft. The whole system was mounted to a customized rack system, Fig. 11.3c.

One sheep was used in this in-vivo study. The sheep was used in the Institutional Animal Care and Use Committee (IACUC) approved study as required by the Animal Welfare Act. To exposure the dura to indentation, a laminectomy was performed on a sheep around the third lumbar vertebra (L3) by removing the lamina (the back part of the vertebra). The indenter was then adjusted toward the dura in a way that the compression force due to dura movement of respiration was minimized. The test was then started. The indenter was moved toward the dura at a speed of 0.05 in/s to a specified displacement. The specified test displacements were 2 mm, 3 mm, and 4 mm. The displacements were deemed as non-discomfortable dura compressions for the sheep.

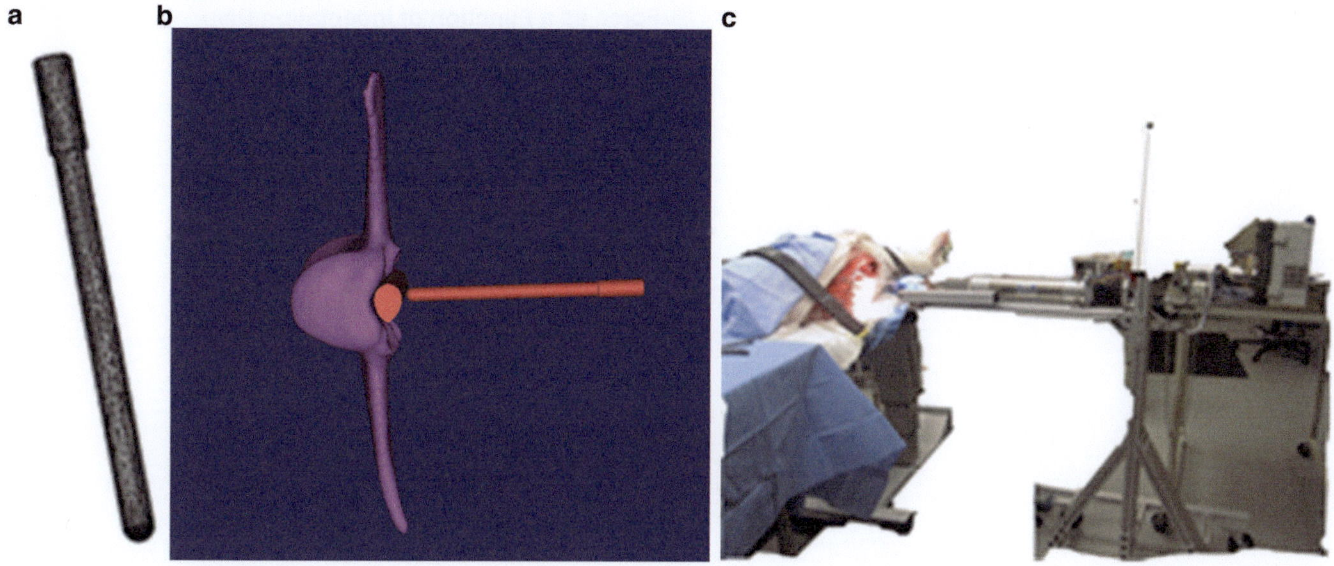

Fig. 11.3 Schematic illustration of indenter and test apparatus

11.2.2 Equivalent Modulus

It is assumed spinal cord with meninges deforms elastically when the indentation displacement is less than 4 mm. The equivalent in-vivo Young's modulus of spinal cord with meninges can be obtained by fitting the measured indentation depth as a function of indenter force using the following model:

$$F = \frac{4}{3}\frac{E}{1-\nu^2}\delta^{\frac{3}{2}}R^{\frac{1}{2}} \tag{11.1}$$

where F is the force applied by the indenter, δ is the indentation depth, R is the indenter diameter, υ is the Poisson's ratio, which is assumed to be 0.45 [8, 9]; E is the equivalent in-vivo Young's modulus of spinal cord with meninges.

11.3 Results and Discussion

Figure 11.4 shows a representative plot of the force as a function of indentation displacement. Table 11.1 provides a summary of the testing conducted in this study. Figure 11.5 shows the Boxplot of the equivalent modulus of spinal cord with meninges at each indentation location. The equivalent modulus has a mean of 0.022 MPa and a standard deviation of 0.014 MPa.

Compared to the elastic modulus of cadaver spinal cord [5–7], 0.26 MPa and the elastic modulus of cadaver dura mater [9], 5 MPa respectively, the equivalent elastic modulus of spinal cord with meninges is much smaller. This can be explained by spinal cord and meninges structure. First, when indenter applies force to the dura mater, the force will be transferred to the CSF area before an equilibrium state is reached. Secondly, the thickness of sheep dura mater is very small, measured as about 0.085 mm with a drop micrometer. This indicates the overall equivalent modulus of spinal cord with meninges can be significantly different from that of individual dura component. Similar data comparison observations have been made for spinal cord with and without pia mater [12].

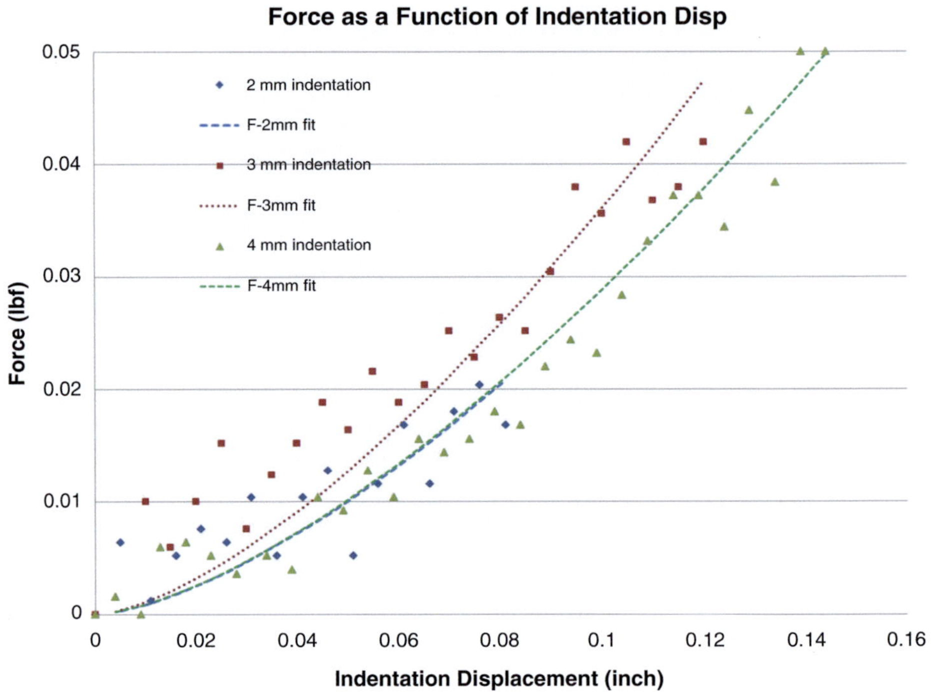

Fig. 11.4 Representative plot of force as a function of indentation displacement

Table 11.1 Indentation testing summary

Test #	Test location	Displacement (mm)	Modulus (MPa)
1	1	2	0.013
2	1	3	0.017
3	1	4	0.013
4	2	4	0.014
5	2	4	0.006
6	2	4	0.006
7	2	4	0.006
8	2	4	0.004
9	3	3	0.048
10	3	3	0.032
11	3	3	0.029
12	3	3	0.028
13	4	3	0.048
14	4	3	0.032
15	4	3	0.029
16	4	3	0.028
17	4	3	0.026

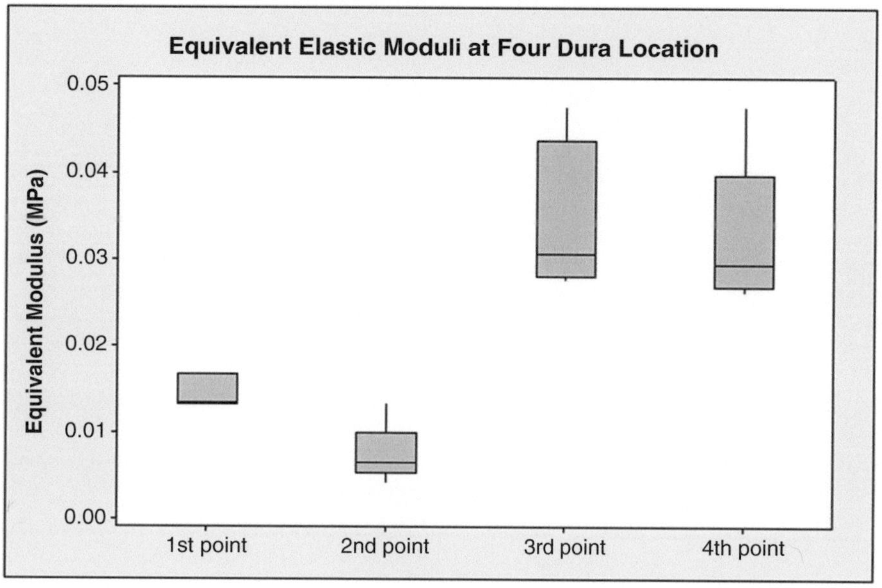

Fig. 11.5 Boxplot of the equivalent modulus at each indentation location

11.4 Summary

- In this study, an in-vivo indentation test method was developed to assess the equivalent Young's Modulus of sheep spinal cord with meninges. The equivalent modulus has a mean of 0.022 MPa and a standard deviation of 0.014 MPa.
- Compared to the elastic modulus of spinal cord and the elastic modulus of dura mater respectively, the equivalent elastic modulus of spinal cord with meninges is much smaller.

References

1. Manchikanti, L., Boswell, M.V., et al.: Comprehensive review of therapeutic interventions in managing chronic spinal pain. Pain Physician **24**, E123–E198 (2009)
2. Barolat, G., Sharan, A.D.: Spinal cord stimulation for chronic pain management. Semin. Neurosurg. **15**(2/3), 151–175 (2004)
3. Cameron, T.: Safety and efficacy of spinal cord stimulation for the treatment of chronic pain: a 20-year literature review. J. Neurosurg. Spine **100**, 254–267 (2004)
4. Manchikanti, A., Boswell, M., Datta, S., Fellows, B., Abdi, S., Vijay, V., Benyamin, R., Falco, F., Helm, S., Hayek, S., Smith, H.: Comprehensive review of therapeutic interventions in managing chronic spinal pain. Pain Physician **12**, E123–E198 (2009)
5. Chang, G., Hung, T., Bleyaert, A., Jannetta, P.: Stress–strain measurement of the spinal cord of puppies and their neurological evaluation. J. Trauma. **21**, 807–810 (1981)
6. Hung, T., Chang, G.: Biomechanical and neurological response of the spinal cord of a puppy to uniaxial tension. J. Biomech. Eng. **103**, 43–47 (1981)
7. Hung, T., Chang, G., Lin, H., Walter, F., Bunegin, L.: Stress–strain relationship of the spinal cord of anesthetized cats. J. Biomech. **14**, 269–276 (1981)
8. Chang, G., Hung, T., Feng, W.: An in-vivo measurement and analysis of viscoelastic properties of the spinal cords of cats. J. Biomech. Eng. **110**, 115–122 (1988)
9. Van Noort, R., Black, M., Martin, T., Meanley, S.: A study of the uniaxial mechanical properties of human dura mater preserved in glycerol. Biomaterials **2**, 41–45 (1981)
10. Breig, A.: Biomechanics of the Central Nervous System. Almzvist & Wiksell, Stockholm (1960)
11. Bilston, L., Thibault, L.: The mechanical properties of the human cervical spinal cord in vitro. Ann. Biomed. Eng. **24**, 67–74 (1995)
12. Mazuchowski, E., Thibault, L.: Biomechanical properties of the human spinal cord and pia mater. Summer Bioengineering Conference, Sonesta Beach Resort in Key Biscayne, Florida, 25–29 June 2003

Chapter 12
Polydimethylsiloxane Lanes Enhance Sarcomere Organization in Human ESC-Derived Cardiomyocytes

Brett N. Napiwocki, Max R. Salick, Randolph S. Ashton, and Wendy C. Crone

Abstract Human embryonic stem cell–derived cardiomyocytes (hESC-CMs) hold great potential in many areas of research such as cardiac tissue regeneration, cardiotoxicity screening and human heart disease models; however, before any of these applications can be realized, hESC-CMs need to progress from an immature phenotype to one that more closely resembles their adult counterparts in vivo. Current immature hESC-CMs can be characterized by their rounded morphology, disorganized contractile apparatus and circumferential gap junction expression. In an effort to improve the maturation of hESC-CMs, prior work in our lab used micropatterned lanes of Matrigel and fibronectin extracellular matrix proteins on glass slides to control cell shape. From these experiments it was found that widths ranging from 30 to 80 μm promoted the best sarcomere development and nuclear alignment in a pure population of hESC-CMs. In this new system, a pure population of hESC-CMs are seeded onto lanes of Matrigel on polydimethylsiloxane (PDMS) to investigate the portability of this technique to other substrate systems and how the substrate stimuli influences maturation. The same trend in nuclear alignment and sarcomere organization with lane width was observed when hESC-CMs were seeded onto lanes of Matrigel on PDMS as was found in the experiments utilizing glass as the substrate. By restricting cell adhesion and controlling substrate stiffness it may be possible to enhance the maturation of hESC-CMs in vitro which will provide a more physiological relevant phenotype that can then be used in the aforementioned applications.

Keywords Cardiomyocyte • Stem cell • Cardiac tissue engineering • Microcontact printing • PDMS

12.1 Introduction

Cardiovascular disease leads to over 800,000 deaths and approximately 610,000 new myocardial infarctions, or heart attacks, every year [1]. Myocardial infarction negatively impacts both the mechanical and electrophysiological properties of the heart, leading to shortness of breath, palpitations, and increased risk of further complications. While heart transplantation is a viable option for patients of end-stage heart failure, limited availability and high invasiveness prevent this from being a common solution for myocardial repair. Clearly, the incentive for the creation of improved models, therapies, and treatments for myocardial repair is extremely high. With the advent of well-established differentiation protocols, cardiomyocytes derived from pluripotent stem cells (hPSC-CMs) appear to be an ideal cell source for myocardial repair. In fact, researchers have shown that hPSC-CMs transplanted into infarcted hearts of mice and guinea pigs were able to integrate with the native myocardium and improve mechanical function [2, 3]; however, in order for greater improvement to occur, the hPSC-CMs must become more mature.

B.N. Napiwocki • R.S. Ashton
Department of Biomedical Engineering, University of Wisconsin-Madison, Madison, WI, USA

M.R. Salick
Materials Science Program, University of Wisconsin-Madison, Madison, WI, USA

W.C. Crone (✉)
Department of Engineering Physics, University of Wisconsin-Madison,
Engineering Research Building, 1500 Engineering Drive, Madison, WI 53706, USA
e-mail: crone@engr.wisc.edu

S.A. Tekalur et al. (eds.), *Mechanics of Biological Systems and Materials, Volume 6*,
Conference Proceedings of the Society for Experimental Mechanics Series, DOI 10.1007/978-3-319-21455-9_12

In vivo, adult cardiomyocytes (CMs) are rod-shaped and contain myofibrils that run parallel to the length of the cell. In contrast, when immature human embryonic stem cell-derived cardiomyocytes (hESC-CMs) are cultured in vitro they are pleomorphic in shape and have no clear myofibril alignment. It is hypothesized that in order to improve the maturation of hESC-CMs, they must resemble their adult counterparts in vivo. hESC-CM maturation is characterized by several phenotypes, including higher calcium conduction speeds and greater contractile strength. In order to control the alignment of hESC-CMs, several research groups have used soft lithography combined with microcontact printing. Recently, our group found a correlation between the micropattern lane width and myofibril organization within a pure population of hESC-CMs [4]. From multiple experiments it was determined that micropatterned lane widths of Matrigel between 30 and 80 μm promoted highly aligned CMs with an organized myofilament structure, but CMs in widths greater than 100 μm never developed aligned myofibrils. In order to produce other characteristics of mature cardiomyocytes, we believe it is necessary to use an underlying substrate that more closely mimics the matrix elasticity of the human heart instead of gold-platted glass slides which were used in the previous experiments.

The elastic modulus of the human heart ranges from 2 to 8 kPa [5]. For the rat myocardium, this value is 10–20 kPa [6]. When neonatal rat ventricular myocytes (NRVMs) were cultured on collagen-coated polyacrylamide (PA) gel substrates of different elastic moduli values ranging from 1 to 50 kPa, the researchers found that sarcomeres assembled best on the 10 kPa stiffness which is similar to the rat native myocardium [7]. In addition to cytoskeletal organization, the overall shape of the NRVMs was uniaxial on the soft and intermediate matrices, 1 kPa and 10 kPa, respectively, while NRVMs on stiff matrices were irregular in shape. These results suggest that hESC-CMs seeded on gels with an elastic modulus similar to the native human myocardium will be rod-shaped and display aligned sarcomeres.

Herein, we describe how Matrigel-coated lanes on polydimethylsiloxane (PDMS) can be used to study the effects of sarcomere development and organization in hESC-CMs. PDMS was chosen because it is optically transparent and the substrate stiffness can easily be modulated by changing the crosslinking ratios of the base and curing agent [8]. While ideally we would like to use PDMS with an elastic modulus similar to the human native myocardium, for proof of concept we mixed the base and curing agent in a 10:1 ratio, suggested by the manufacturer, which results in an elastic modulus of 1600 kPa (data not shown).

12.2 Methods

12.2.1 hESC Culture

Cardiomyocytes were derived from an engineered H9 hESC line containing a cardiac troponin T GFP (cTnT-GFP) promoter. This cTnT-GFP promoter line exhibits resistance to the antibiotic Zeocin along with cardiac troponin T expression indicated by the presence of GFP, allowing for the purification of cTnT-expressing cells. hESCs were maintained on vitronectin-coated polystyrene plates and fed with E8 media (replaced daily) prior to differentiation [9]. E8 media was made using DMEM/F-12 (Life Technologies), L-ASCORBIC acide 2-phosphate sesquimagnesium salt hydrate, sodium selenite, sodium bicarbonate, holotransferrin, insulin, FGF-2 (WiCell), and TGF-β1 (Peprotech). Unless otherwise noted, all reagents were purchased from Sigma-Aldrich. Every 4 days the hESCs were passaged at a 1:12 ratio using versene (Life Technologies) and were seeded with ROCK inhibitor (Tocris).

12.2.2 hESC Cardiomyocyte Differentiation

To produce cardiomyocytes from the cTnT-promoting H9 hESC line (hESC-CMs), a modified version of the small molecule Wnt-agonist method of differentiation was used [10]. Briefly, on day-3 the cTnT-GFP hESCs were seeded onto Matrigel-coated (BD Biosciences) 12-well plates at a density of 400,000 cells per well and fed with E8 medium supplemented with 5 μm ROCK inhibitor (Tocris). On day-1, the cells were again fed with E8 medium. On day 0 (exactly 72 h after initial seed), cells were fed with RPMI (Life Technologies) supplemented with B27-insulin (Life Technologies), 10.5 μm of the Wnt agonist CHIR 99021 (Tocris) and 1 μg/mL of insulin (Sigma). On day 1, cells were fed with RPMI supplemented with B27-insulin. On day 3, cells were treated with RPMI supplemented with B27-insulin and 2.5 μm of the Wnt inhibitor IWP4 (StemGent). On day 5, cells were fed with RPMI containing B27-insulin and then treated with RPMI containing B27-complete on days 7, 8, 9, and every 3 days thereafter. Upon observation of contractions (typically between days 9–12),

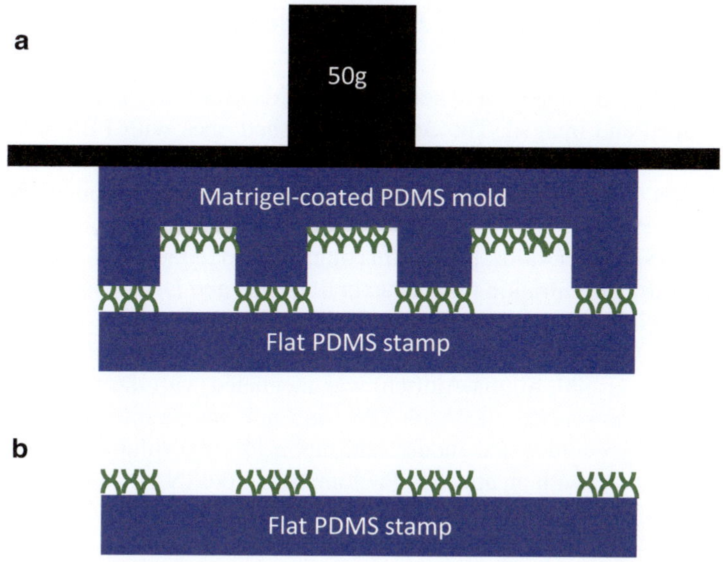

Fig. 12.1 A schematic representation of microcontact printing. (**a**) The Matrigel-coated PDMS mold was placed on top of a flat PDMS stamp. A glass slide and 50 g weight were added on top of the Matrigel-coated PDMS mold to insure uniform pressure distribution. (**b**) The following day the Matrigel-coated PDMS mold was removed and Matrigel lanes were transferred from the PDMS mold to the flat PDMS stamp. Both schematics are shown from a side view and 'x' represents Matrigel

Zeocin treatments were applied in which 100 μg/mL per day was added to the RPMI-B27-complete medium for 3 consecutive days to remove non-cTnT expressing cell types. The resulting cells were confirmed by flow cytometry to be ~98 % cardiomyocytes.

12.2.3 Microcontact Printing

The patterning technique in this study utilized microcontact printing to generate Matrigel-coated PDMS substrates upon which hESC-CMs could adhere. Several steps were required to make the PDMS mold. First, a template with the feature designs (lanes with varying widths) was created in AutoCad (autodesk) and sent to the Stanford Microfluidics Foundry for fabrication of a photomask and a 4-in. patterned Si wafer (Stanford University, Santa Clara County, CA). Using soft photolithography techniques, the Si wafer was spin coated with an SU-8 negative photoresist, and exposed to UV light. This design incorporated long lanes with widths ranging from 40 to 240 μm (with each subsequent lane increasing by 20 μm until 240 μm was reached, and the pattern then repeated itself), which provided for an efficient seeding surface. Poly (dimethyl siloxane) (PDMS) (Sylgard 184 Kit) was mixed in a 10:1 base to crosslinking ratio and then poured over the silicon master and cured overnight at 60 °C. The following day the patterned PDMS mold was removed from the Si master and coated with Matrigel (83 μg/mL) (WiCell) for 24 h. After 24 h, the Matrigel-coated PDMS mold was dried with N_2 and placed on top of a flat PDMS stamp. A glass slide and 50 g weight were added on top of the two PDMS stamps to ensure even pattern transfer from the Matrigel-coated PDMS mold to the flat PDMS stamp (Fig. 12.1a). This setup was incubated overnight at 37 °C. Afterwards, the flat PDMS stamp now containing lanes of Matrigel was removed from the Matrigel-coated PDMS mold and seeded with hESC-CMs (Fig. 12.1b).

12.2.4 hESC-CM Seeding and Imaging

Purified hESC-CMs were seeded onto the flat PDMS substrates 7 days after finishing the Zeocin purification step. The purified hESC-CMs were then washed 3 times with PBS to help remove debris. Cells were dissociated from the original culture via exposure to TrypLE (Life Technologies) for 15–20 min. Plates were agitated every 5 min to help break up significant cell-cell binding and extracellular matrix (ECM) deposits formed during the differentiation culture. Cells were then added to flat PDMS substrate containing lanes of Matrigel. The hESC-CMs were seeded at a density of 30,000 cells/cm^2 which allows for confluent cell attachment while avoiding multiple cells from piling on top of one another. The media was replaced with fresh RPMI-B27-complete every day thereafter.

12.2.5 Imaging and Immunofluorescence

Bright field images were acquired every day for 5 days after the original seeding. After 5 days of culture the samples were stained and imaged. The cells were washed once with PBS and then exposed to 4 % paraformaldehyde (PFA) (Electron Microscopy Sciences) for 15-min at room temperature. The cells were washed again with PBS and then treated with 0.1 % Triton (Sigma) for 6 min at room temperature to permeabilize the cell membrane. Again, the cells were washed with PBS and treated for 30 min with a blocking solution consisting of PBS, 2 % FBS, 0.1 % Triton, 11.2 mg/mL glycine, and 50 mg/mL BSA. Afterwards, phalloidin conjugated to tetramethylrhodamine B isothiocyanate (TRITC) (Sigma) was applied at a 50 mg/mL concentration to label actin filaments and DAPI was applied at a 1:1000 dilution to label nuclei. Cells were washed with PBS and transferred to coverslips, where they were mounted using ProLong Gold Antifade (Life Technologies). To visualize the transfer of Matrigel from the patterned PDMS mold to the flat PDMS stamp, the primary antibody rabbit anti-laminin (1:500, Sigma-Aldrich) was incubated with the flat PDMS stamp overnight at 4 °C. The next day the flat PDMS stamp was washed once with PBS for 5 min and then treated for 1 h with blocking solution containing the secondary antibody, Cy3-conjugated mouse anti-rabbit (1:1000 dilution, Abcam). Samples were imaged on a Nikon A1RSi Confocal Microscope with an attached Photometrics CoolSNAP HQ2 camera.

12.2.6 Nuclear Alignment Measurement

The alignment of cardiomyocytes was determined via the measurement of nuclear alignment, as described in our prior publication [4]. Using the Nikon Elements—AR software, ROIs were automatically determined in the DAPI (nuclear) channel of the confocal images. Overlapping nuclei were manually separated to keep them from being quantified as a single nucleus. The major axis of the nuclei was then determined automatically, and the angle of this axis was used to determine cell alignment. Cells were then quantified as "aligned" if their major axis fell within 20° of the feature axis on which the cell was adhered.

12.3 Results and Discussion

12.3.1 Immunofluorescence of Matrigel Transfer

An anti-laminin antibody was used to visualize the absorption of Matrigel onto the flat PDMS stamp, since laminin is one of the ECM proteins that make up Matrigel [11]. As can be seen in Fig. 12.2, some of the Matrigel lanes transferred well to the flat PDMS stamp while other areas did not have distinct edges (denoted by white arrows). The regions that did not transfer well could have occurred during the removal of the Matrigel-coated PDMS mold from the flat PDMS stamp or from excess Matrigel being pushed out under the pressure of the 50 g weight. Regions like this are problematic when analyzing nuclear and sarcomere alignment because it is believed that the lane boundary plays an important role in determining cell shape and cytoskeleton organization; therefore, as mentioned below, only regions with distinct edges were used when measuring nuclear alignment.

12.3.2 hESC-CM Attachment and Contraction on Lanes of Matrigel on PDMS

Bright field images were taken every day during the 5 day culture period to observe hESC-CM spreading and contraction (Fig. 12.3). Some of the hESC-CMs attached to the lanes of Matrigel on PDMS after 24 h and began to spread while others exhibited a rounded morphology. Further cell spreading was observed as the hESC-CMs were cultured for longer time periods but migration was not significant. Several rounded hESC-CMs had attached to regions between Matrigel lanes 1 day after the initial seeding, but most of these cells were removed during media changes. Small contractions were detected 1 day after the initial seeding and increased in strength during the extended culture period. If a cardiomyocyte bridged two adjacent lanes, then the two lanes of hESC-CMs contracted synchronously. Otherwise, if not attached, the two lanes would contract at different times.

Fig. 12.2 Matrigel (*red*) lanes on a flat PDMS substrate visualized with an anti-laminin antibody. *White arrows* indicate regions where the Matrigel did not show perfect fidelity in the transfer from the Matrigel-coated PDMS mold. Scale bar 100 μm

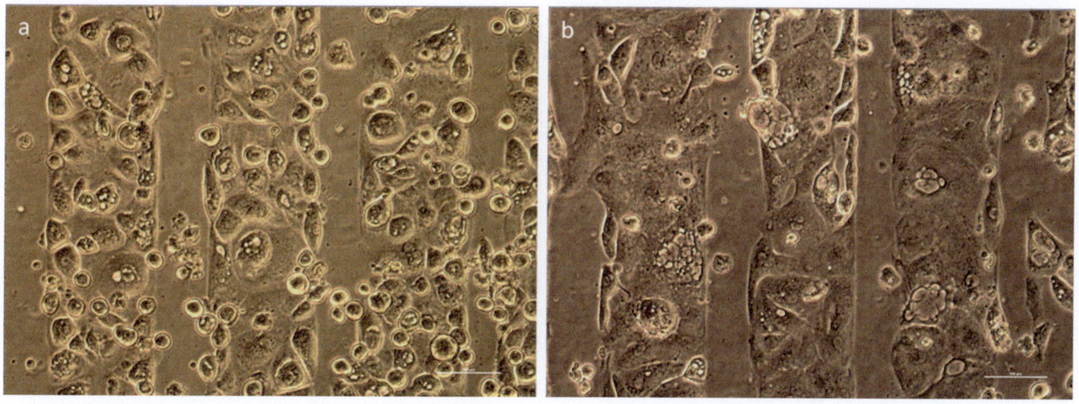

Fig. 12.3 hESC-CMs adhered to lanes of Matrigel on a flat PDMS substrate. (**a**) Bright field images taken 1 day and (**b**) 5 days after the initial seed. Both images are taken of the same location. Scale bar 100 μm

12.3.3 Sarcomere Organization in hESC-CMs Adhered to PDMS

Sarcomere organization was analyzed qualitatively by viewing phalloidin confocal images of hESC-CMs seeded on different lane widths of Matrigel on PDMS (Fig. 12.4). The most aligned sarcomeres were present in hESC-CMs seeded on lane widths of 38.5 μm. This width also produced the most elongated hESC-CMs. In larger lane widths, such as 78.5 μm, the hESC-CMs on the edges had aligned cytoskeletal networks with respect to the feature direction while the hESC-CMs in the middle of the feature were less aligned. As the feature width increased past 100 μm, the hESC-CMs in the middle of the lanes displayed no clear sarcomere organization. These observations are consistent with those where hESC-CMs were seeded onto micropatterned gold slides [4].

12.3.4 Nuclear Alignment of hESC-CMs

Nuclear alignment of hESC-CMs was determined quantitatively using Nikon Elements—AR software. As previously described, cells were categorized as "aligned" if their major axis was within 20° of the feature axis to which they were adhered. As can be seen in Fig. 12.5, the nuclear alignment followed the same trend as the sarcomere organization. The best nuclear alignment on PDMS was observed in lane widths of 40–100 μm. In lane widths larger than 100 μm, the amount of

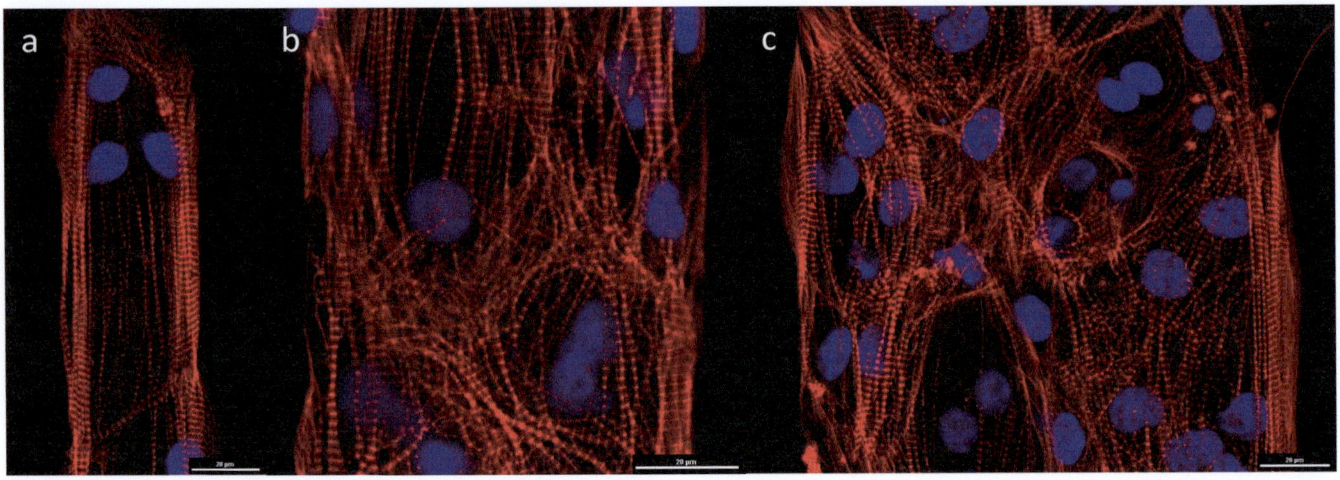

Fig. 12.4 Sarcomere organization in hESC-CMs seeded on different lane widths of Matrigel on PDMS, including (**a**) 38.5 μm, (**b**) 78.5 μm and (**c**) 138.5 μm. Scale bar is 20 μm

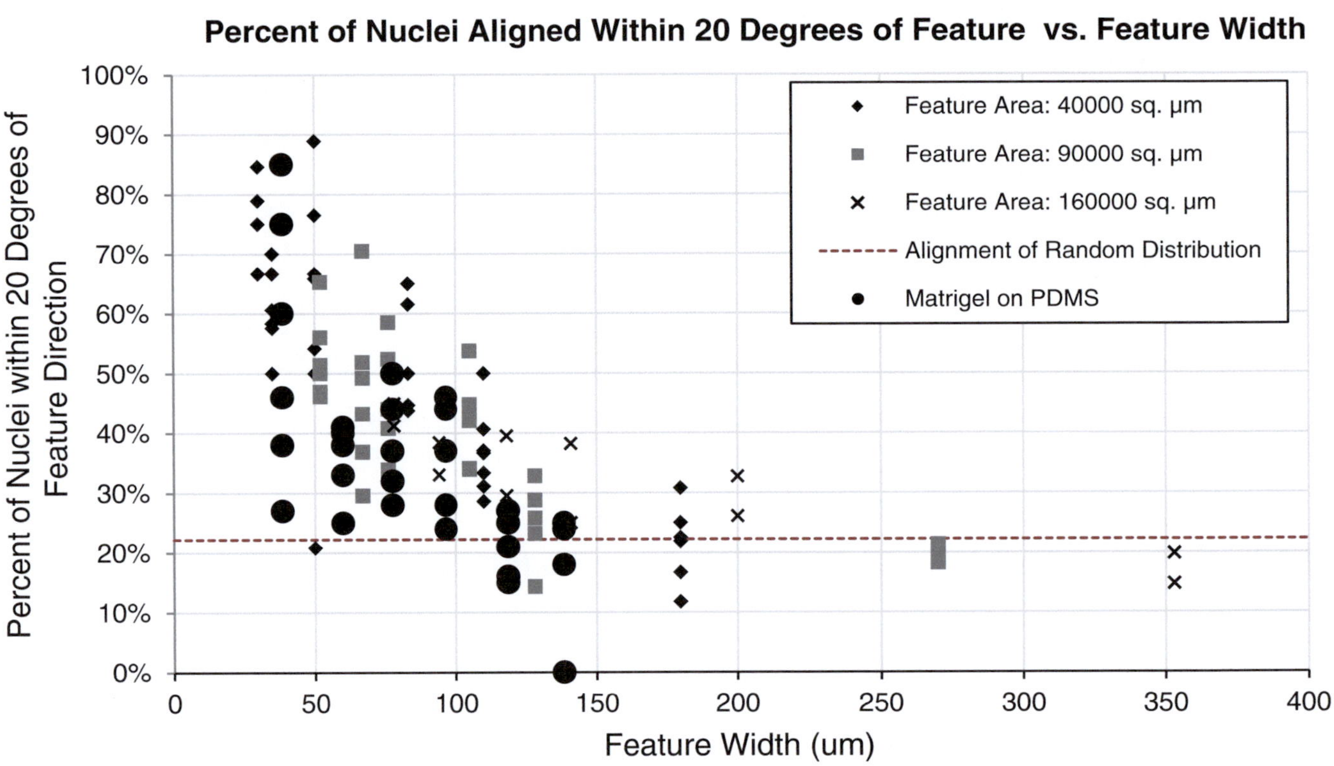

Fig. 12.5 Percentage of nuclei on different feature widths that are aligned with respect to feature axis. Results described in this paper shown with (*filled circles*) and compared to those of Salick et al. [4] using the same methods on glass substrates

nuclei that were aligned was close to 22 % which is the expected value of cells that have no directionality present. The results obtained in this study using PDMS as the underlying substrate were overlaid with our previous published results using Matrigel lanes on micropatterned gold slides [4]. The nuclear alignment followed the same trend irrespective of the underlying substrate. It should be noted that three of the data points for the 38.5 μm feature width in Fig. 12.5 fell below 50 % of the nuclei being aligned. This may have occurred because the edges of the Matrigel lanes on PDMS were occasionally less distinct than on glass slides. In the previous system using glass slides, Matrigel lanes are surrounded by non-adherent poly (ethylene glycol) (PEG) regions which are inert to protein absorption. In this current system utilizing PDMS as the substrate,

Matrigel lanes are bordered by bare PDMS regions that are capable of Matrigel absorption. Since the ECM boundary conditions control cytoskeletal organization, it follows that lanes with less distinct edges will have a less organized sarcomere structure and their nuclei won't be as aligned as cells on lanes with smooth edges.

12.4 Conclusion

The alignment of cardiomyocytes in vitro is essential for their maturation. It has been shown previously that the underlying substrate stiffness can affect the morphology of the cardiomyocytes [7]. Additionally, micropatterning techniques have been used to force alignment in a pure population of hESC-CMs [4]. In this study, it was demonstrated that the lane technology used to promote alignment of hESC-CMs on micropatterned gold slides can be translated to different substrates, such as PDMS, and produce the same sarcomere organization and nuclear alignment results. Future work will focus on ways to transfer Matrigel without having regions with less defined edges. In conclusion, by combining multiple cues (i.e., substrate stiffness and topographical features) it may be possible to create a more physiologically-relevant model of human heart tissue which will aid in regenerative medicine therapies, disease modeling, drug testing, developmental and cardiotoxicity studies.

Acknowledgments This research was supported with funds from the National Institutes of Health Grant K18 HL105504 from the Heart, Blood and Lung Institute and the Graduate School of the University of Wisconsin-Madison.

References

1. Jiaquan, X., et al.: Deaths: final data for 2007. Natl. Vital Stat. Rep. **58**(19), 1–136 (2010)
2. van Laake, L.W., et al.: Human embryonic stem cell-derived cardiomyocytes survive and mature in the mouse heart and transiently improve function after myocardial infarction. Stem Cell Res. **1**(1), 9–24 (2007)
3. Shiba, Y., et al.: Human ES-cell-derived cardiomyocytes electrically couple and suppress arrhythmias in injured hearts. Nature **489**(7415), 322–325 (2012)
4. Salick, M.R., et al.: Micropattern width dependent sarcomere development in human ESC-derived cardiomyocytes. Biomaterials **35**(15), 4454–4464 (2014)
5. Wen, H., et al.: Magnetic resonance imaging assessment of myocardial elastic modulus and viscosity using displacement imaging and phase-contrast velocity mapping. Magn. Reson. Med. **54**(3), 538–548 (2005)
6. Berry, M.F., et al.: Mesenchymal stem cell injection after myocardial infarction improves myocardial compliance. Am. J. Physiol. Heart Circ. Physiol. **290**(6), H2196–H2203 (2006)
7. Jacot, J.G., McCulloch, A.D., Omens, J.H.: Substrate stiffness affects the functional maturation of neonatal rat ventricular myocytes. Biophys. J. **95**(7), 3479–3487 (2008)
8. Palchesko, R.N., et al.: Development of polydimethylsiloxane substrates with tunable elastic modulus to study cell mechanobiology in muscle and nerve. PLoS One **7**(12), e51499 (2012)
9. Chen, G., et al.: Chemically defined conditions for human iPSC derivation and culture. Nat. Methods **8**(5), 424–429 (2011)
10. Lian, X., et al.: Robust cardiomyocyte differentiation from human pluripotent stem cells via temporal modulation of canonical Wnt signaling. Proc. Natl. Acad. Sci. **109**(27), E1848–E1857 (2012)
11. Kleinman, H.K., Martin, G.R.: Matrigel: basement membrane matrix with biological activity. Semin. Cancer Biol. **15**(5), 378–386 (2005)

Chapter 13
Electrospin Fiber Affect on the Strength of Metal–Cement Interfaces

Shahram Riahinezhad, Yanling Li, and Morshed Khandaker

Abstract The research objectives were to (1) determine electrospun polymer fiber adhesion on Titanium (Ti) using qualitative wear tests; (2) determine the effect of fiber material viscosity on the surface coating of Ti; and (3) determine bonding strength between Ti/ Poly-Methyl-Methacrylate (PMMA) under static load. Polycaprolactone (PCL)-Acetone and PCL-PMMA-Acetone fibers were produced on Ti using electrospinning process. Fiber coated Ti surfaces were scratched using sharp edge needle to evaluate the fiber stickiness to the Ti surface. PCL and PMMA were each mixed with acetone using a sonicator, and then stirred together. Rheometer was used to determine viscosity of PCL-PMMA-Acetone mixtures. Qualitative adhesion tests showed that PCL-PMMA-Acetone solution had greater stickiness compared to PCL-Acetone. Low viscosity of 0.158 ± 0.048 Pa s was achieved for producing PCL-PMMA-Acetone fiber which had averaged size of 1.79 ± 0.4 μm. This study found pull out interface fracture shear strength of PCL-Acetone (1.06 ± 0.23 MPa, n = 3) and PCL-PMMA-Acetone (0.57 ± 0.18 MPa, n = 4) fiber coated implant-PMMA cement interfaces were significantly higher compare to uncoated Ti-PMMA interfaces (0.34 ± 0.04 MPa, n = 3).

Keywords Titanium • Cement • Interface • PMMA • Polycaprolactone • Fracture toughness • Implant

13.1 Background

The big challenges of orthopedic research are the implant loosening or the breakage at the implantation sites and improve the bonding between the interface materials [1]. Increase the bounding strength between Ti and PMMA by improving the surface properties of implant was the main motivation of this research [2]. Electrospinning technology is used to fabricate fiber with different morphologies and multidimensional structures and these fiber are widely applied in medical diagnosis, tissue engineering, replica molding and other applications. Poly-caprolactone (PCL) is one kind of synthetic polyesters and compatible with many types of polymers. It is bio-compatible, cheap, easy to process. Thus PCL is being considered as bio-materials for bone tissue, nerve tissue and drug delivery systems [3]. Poly (methyl methacrylate) (PMMA) is known for its good mechanical as bone cement material. Besides, PMMA is also used as lenses, bone substitutes and drug delivery systems. It is also being tested as a fiber material to achieve long-term mechanical stability after implantation. PCL-only and PCL-PMMA fiber were produced and studied in this research. The effect of electrospun material on the fracture strength of Ti/PMMA interface is not known and will be understood in our research [4]. An nanofiber-coated implant may reduces implant failures due to higher Osseo-integration. Failing implants can cost a patient much more due to the cost of revision surgery. Therefore, fiber-coated implant may economically beneficial to patients.

13.2 Materials and Methods

Titanium (Ti) bars (6Al-4 V ELI, ASTM B 348 standard, grade 23, biocompatible) of dimension (22 × 12 × 2) mm were purchased from Titanium Metal Supply, Inc., Poway, CA. Cobalt™ HV bone cement (Biomet Inc., Warsaw, IN) was used as the polymethyl methacrylate (PMMA) cement. PCL was purchased from Sigma Aldrich. PCL was selected as fiber material

S. Riahinezhad • Y. Li • M. Khandaker (✉)
Department of Engineering and Physics, University of Central Oklahoma, Edmond, OK 73034, USA
e-mail: mkhandaker@uco.edu

S.A. Tekalur et al. (eds.), *Mechanics of Biological Systems and Materials, Volume 6*,
Conference Proceedings of the Society for Experimental Mechanics Series, DOI 10.1007/978-3-319-21455-9_13

since it is biocompatibility, biodegradable and electrostatic. PCL solution was prepared by ultrasonic (Sonics & Materials, Inc., Vibra-cell VCX 130) mixing of 7.69 wt% of PCL beads with acetone. The sonication process was carried out at approximately 80 °C for an hour. The solution was poured into a glass syringe in an infusion pump (Harvard Ins.). The cylindrical titanium implant were coated by PCL and PCL-PMMA fiber using the electrospinning process. This process was chosen because of its capability of high precision controlling deposition of the fibers and producing nano-level fibers. The titanium implant was spun at high speed with a DC motor which is used in conjunction with a Proportional Integral Derivative (PID) control system to control the revolutions of the motor under the electrospinning setup. The motor was mounted on Newport precision linear stage to provide translational movement to the titanium along with the rotation.

The entire experimental process used in this study is shown in Fig. 13.1. Four PMMA-acetone and PCL-acetone polymer solutions compositions (1.5 w/v %, 2 w/v %, 2.5 w/v %, and 5 w/v %) were mixed to prepare different viscous solutions. PCL and PMMA were each mixed with acetone using the Sonicator for 45 min and 2 h, and PMMA with acetone was stirred for 45 min after sonication. After two solutions were prepared, two solutions were blended together for 30 min [3]. A Bohlin CVO Rheometer was used to determine viscosity of different PCL-PMMA-Acetone mixture.

Two synthesized electrospun materials, PCL-Acetone, and PCL-PMMA-Acetone were selected to conduct the qualitative adhesion tests on a Ti surface using Buehler® MetaServ 250 Grinder-polisher. Sonics Vibra-Cell Sonicator and Corning® stirring hot plates were used for mixing the solution.

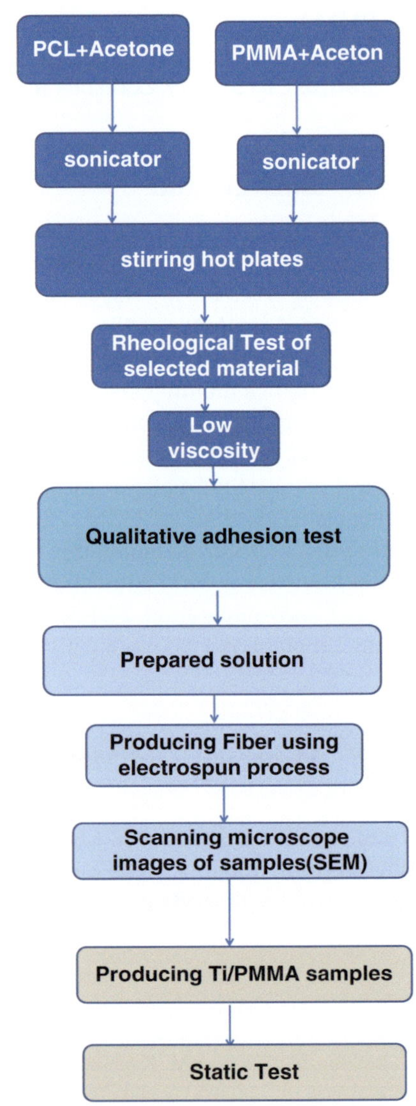

Fig. 13.1 Electrospin setup for the deposition of aligned fiber on titanium rod

For creating fibers from a PCL liquid polymer solution, PCL solution were placed in a 5 mL syringe. During the electrospinnig process, feeding rate and distance between needle and collector were at 1–2 mL and 10–15 cm, respectively [4]. Upon applying voltage 3–15 kV (DC) using a high-voltage power supply (Gamma high voltage research), a fluid jet was ejected from the tip of the needle.

To prepare PCL-PMMA soluntion, PCL and PMMA were each independently mixed with acetone using a sonicator, and then stirred together. Rheometer was used to determine viscosity of PCLPMMA-Acetone mixtures to find the appropriate viscosity to produce fiber using electrospin process.

Each of the fiber coated Ti surfaces were scratched using sharp ended needle to evaluate the fiber stickiness to the Ti surface.

A 3″ length × 3/8″ diameter titanium Ti rod was used for the Ti/PMMA model. With the weight to volume ration of 2:1, 1.23 g of PMMA power was well mixed with 0.613 mL Methyl Methacrylate (MMA) monomer. A custom made setup was used to pot the Ti rod by 5 mm thickness and 10 mm height PMMA cement layer in a 3D printed mold as shown in Fig. 13.2. The Ti/PMMA sample was placed in the Test Resources 800LE4 universal testing machine to be put under tension loading under quasi-static condition. The breaking force over the contact surface area between titanium and cement were used to calculate the interface fracture shear strength.

13.3 Results

Qualitative adhesion tests showed that PCL-PMMA-Acetone polymer material had greater stickiness. By mixing PCL and PMMA with different compositions, composition of **0.15 g** of **PMMA** and **0.5 g** of **PCL** each mixed with **10 mL acetone** was able to produce fine fiber with a low viscosity of **0.157 ± 0.048 Pa s** as shown in Table 13.1.

Fiber distribution was measured using Hitachi TM 3000 Scanning electron microscope (SEM) images (Fig. 13.3). The results (Table 13.2) shows that the ranges of diameter of fibers and distances between two adjacent fibers are mainly micrometer ranges.

Qualitative adhesion tests showed that PCL-PMMA-Acetone solution had greater stickiness compared to PCL-Acetone. Low viscosity of 0.158 ± 0.048 Pa s was achieved for producing PCL-PMMA-Acetone fiber which had averaged size of 1.79 ± 0.4 μm.

Compare of the bonding strength of two electrospun fiber coated Ti/PMMA interfaces under static loads (the strength load between titanium rods with PMMA, and titanium rods with fibers and PMMA) were done. The fracture load was plotted as shown in Fig. 13.4. This study found pull out interface fracture shear strength of PCL-Acetone (1.06 ± 0.23 MPa, n = 3) and PCL-PMMA-Acetone (0.57 ± 0.18 MPa, n = 4) fiber coated implant-PMMA cement interfaces were significantly higher compare to uncoated Ti-PMMA interfaces (0.34 ± 0.04 MPa, n = 3).

Fatigue experiments will be conducted for Ti/PMMA models with and without fibers coating using Test Resources universal testing machine in future. Further wear and biological tests are required for fiber coated implants use in orthopedic or orthodontic applications.

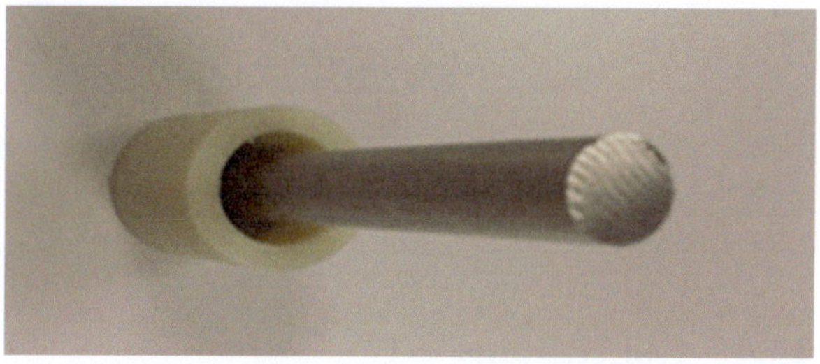

Fig. 13.2 Fabricated Ti/PMMA sample for static and fatigue tests

Table 13.1 Rheological results of PCLPMMA-Acetone mixture

Trials	1	2	3	4	Mean
PCL-PMMA (1.5 w/v %)	0.13	0.23	0.14	0.13	0.1575

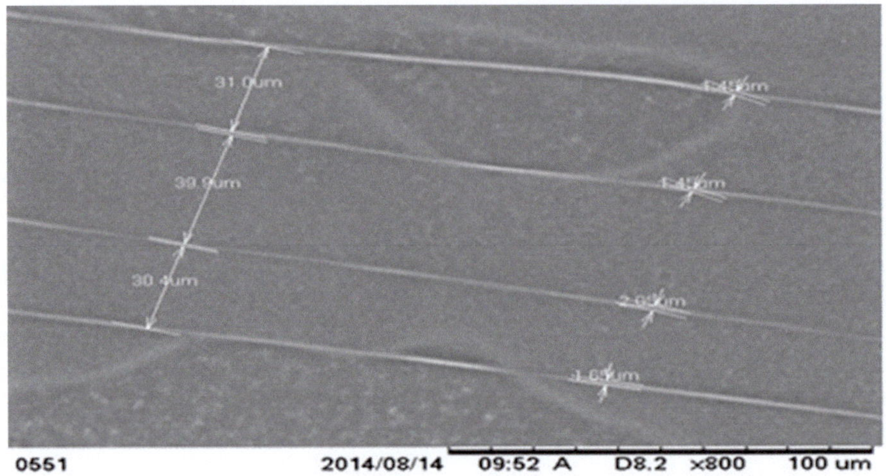

Fig. 13.3 Scanning electron microscope images of samples

Table 13.2 Produced fiber distribution

	Sample		
Description	1	2	3
Number of fibers intersects 35 μm length line (PCL-Acetone)	6	5	6
Range of diameter of fiber (PCL-Acetone)	1.07 μm to 410 nm	1.05 μm to 1.56 μm	1.07 μm to 7.28 μm
Distance between two fibers (PCL-Acetone)	4.54 μm to 7.26 μm	7.26 μm to 9.68 μm	4.57 μm to 7.26 μm

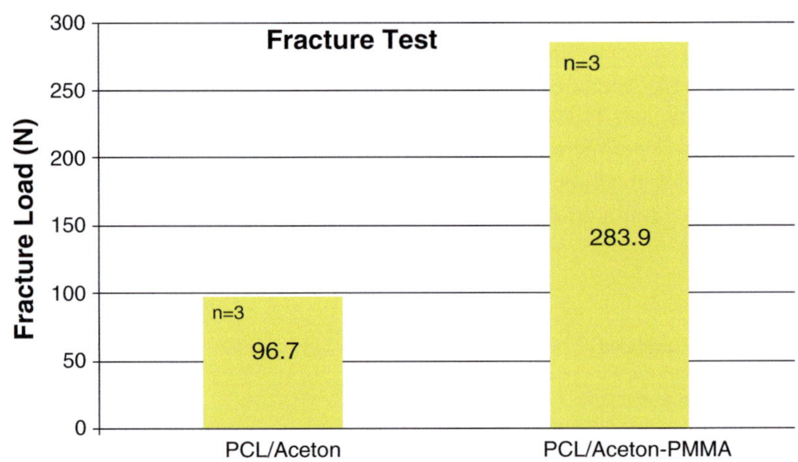

Fig. 13.4 Measure the interface fracture strength of Ti/PMMA samples

13.4 Conclusions

The goal of this research is an efficient bond interface between the Ti and PMMA by applying nanofibers to the surface of the implant through an electrospinning process. Objectives in this research were to: (1) measure the fiber material viscosity, (2) to determine the effect of electrospun polymer fiber adhesion on Ti, and (3) to compare the bonding strength of two electrospun fiber coated Ti/cement interfaces under static loads. An uncoated and fiber coated cylindrical model was tested under static structural analysis. Our experimental study on cylindrical model found significant increase of pull out static strength for fiber coated implant compare to uncoated implant. During this study, the fiber material viscosity of PCL-Acetone,

PMMA-Acetone-PCL were measured. The effect of electrospinning polymer fiber adhesion on Ti and check the fiber stickiness on the titanium rod were determined and compared. Under static conditions, Ti/PMMA model with and without different polymeric fibers were tested to determine the fracture strength. The results shows that the adhesion and strength of fiber is controlled by the type of the electrospin material.

Acknowledgements This research was made possible by grant 8P20GM103447 from the US National Institutes of Health and an on-campus faculty grant program from the University of Central Oklahoma Office of Research and Grants.

References

1. Lu, W., Sun, J., Jiang, X.: Recent advances in electrospinning technology and biomedical applications of electrospun fibers. J. Mater. Chem. B **2**, 2368–2380 (2014)
2. FDA U.S. Food and Drug Administration. Retrieved from http://www.fda.gov/medicaldevices/products andmedicalprocedures/implantsandprosthetics/metalonmetalhipimplants/ucm241770.htm. Accessed 4 Jul 2014
3. Tillman, B.W., Yazdani, S.K., Lee, S.J., Geary, R.L., Atala, A., Yoo, J.J.: The in vivo stability of electrospun polycaprolactone-collagen scaffolds in vascular reconstruction. Biomaterials **30**, 583–8 (2009)
4. Khandaker, M., Utasha, K.C., Morris, T.: Interfacial fracture toughness of titanium-cement interfaces: effects of fibers and loading angles. Int. J. Nanomedicine **9**(1), 1689–1697 (2014)

Chapter 14
A Nanomechanics Based Investigation into Interface Thermomechanics of Collagen and Chitin Based Biomaterials

Tao Qu, Devendra Verma, and Vikas Tomar

Abstract From the biological/chemical perspective, interface concepts related to cell surface/synthetic biomaterial interface and extracellular matrix/biomolecule interface have wide applications in medical and biological technology. Interfaces control biological reactions, and provide unique organic microenvironments that can enhance specific affinities, as well as self-assembly in the interface plane that can be used to orient and space molecules with precision. Interfaces also play a significant role in determining structural integrity and mechanical creep and strength properties of biomaterials. Structural arrangement of interfaces combined with interfacial interaction between organic and inorganic phases significantly affects the mechanical properties of biological materials, allowing in particular for a unique combination of seemingly "in-consistent" properties, such as fracture strength and tensile strength being both high—as opposed to traditional engineering materials, which have high fracture strength linked to low tensile strength and vice-versa. This work presents a framework to understand this correlation by presenting a quantified information regarding the effect of interfaces on overall mechanical deformation of two widely simulated materials systems based on Collagen-Hydroxyapatite and Chitin-Calcite interfaces. Analyses point out specific role of interface chemistries in the effect the interfaces have on overall structural mechanical properties.

Keywords Biological interfaces • Stress/strain relationship • Collagen • Chitin • Interface viscosity • Interface mechanics

Biological materials have evolved over millions of years and are often found as complex composites with superior properties compared to their relatively weak original constituents. Such materials often combine two properties which are contradictory, but essential for their functioning. A unique feature that determines their properties, is interfacial interaction between organic and inorganic phases in the form of protein [e.g., chitin (CHI) or tropocollagen (TC)]-mineral [e.g., calcite (CAL) or hydroxyapatite (HAP)] interfaces.

In the structural studies of such biological materials, it is observed that at the mesoscale (~100 nm to few μm), the mineral crystals are preferentially aligned along the length of the organic phase polypeptide molecules in a hierarchical (e.g., staggered or Bouligand pattern) arrangement, Fig. 14.1 [1–5]. Interfaces are perceived to play a significant role in the stress transfer and the consequent improvements in stiffness and strength of such material systems. However, how exactly the change in interfacial chemical configuration implies change in mechanical properties in such materials, is a big subject of debate. One of the most important aspects of understanding the influence of interfaces on natural material properties, is the knowledge of how stress transfer occurs across the organic–inorganic interfaces. The multi-component hierarchical structure of biomaterials results in organic–inorganic interfaces appearing at different length scales, i.e., between the basic components at the nano-scale, between the mineralized fibrils at the micro-scale, and between the layers of the multi-layered structures at micro- or macro-scale. Earlier studies have focused on the role of interface related mechanisms in determining overall mechanical deformation properties, the real aspect of stresses at interfaces while the mechanical deformation is going on, still remains unaddressed. In this context, important questions are: For a given peak tensile strength of a given material, which position of total strength is attributed to interface strength? What is the contribution of interface sliding in time dependent deformation observed in a simple tension test of a given material sample? Recently, simulation performed by Qu and Tomar [6] have pointed out some important aspects in answering such questions.

T. Qu • D. Verma • V. Tomar (✉)
School of Aeronautics and Astronautics, Purdue University, West Lafayette, IN 47907, USA
e-mail: tomar@purdue.edu

S.A. Tekalur et al. (eds.), *Mechanics of Biological Systems and Materials, Volume 6*,
Conference Proceedings of the Society for Experimental Mechanics Series, DOI 10.1007/978-3-319-21455-9_14

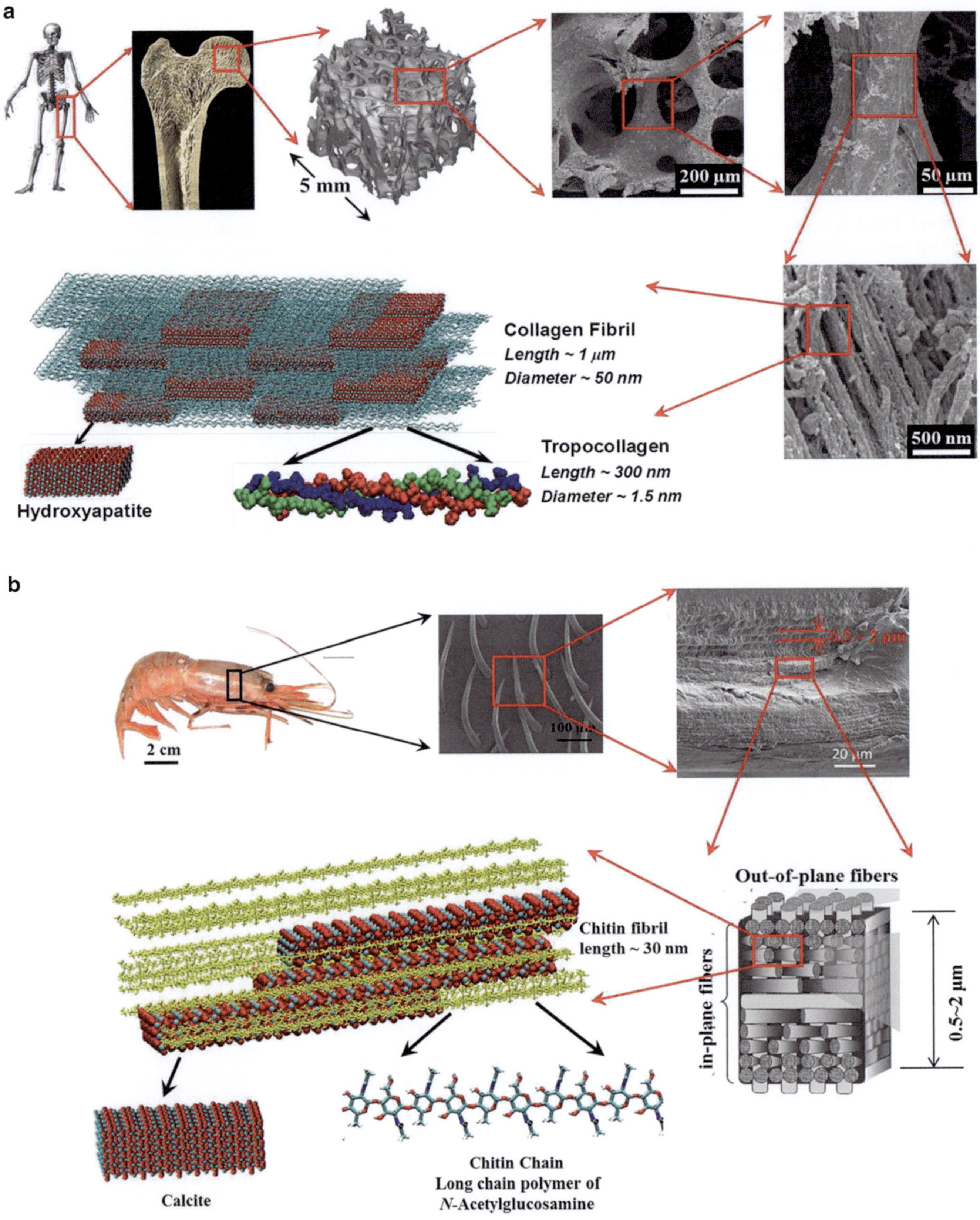

Fig. 14.1 A schematic of the hierarchical structure of (**a**) bone and (**b**) shrimp exoskeleton

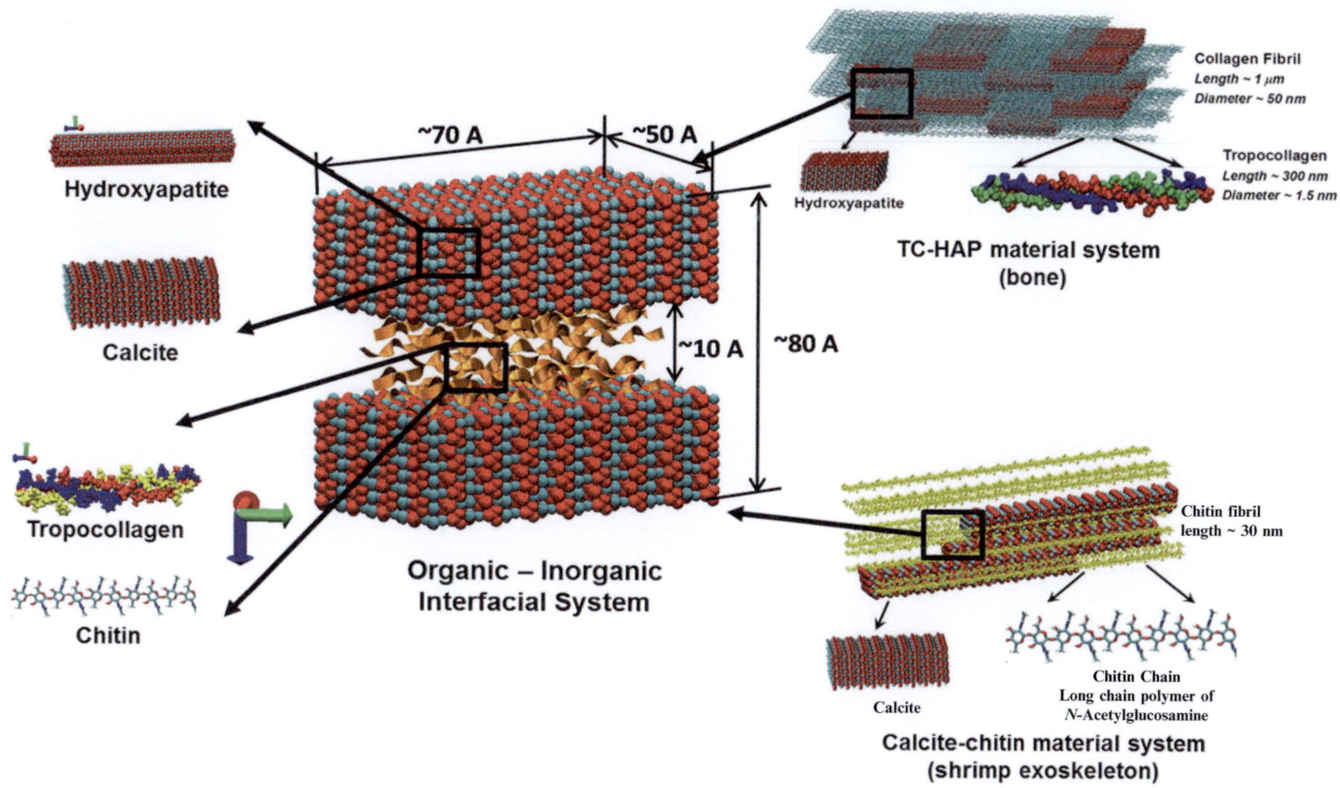

Fig. 14.2 A schematic showing configuration of the analyzed TC-HAP and CHI-CAL interfaces

Figure 14.2 shows a schematic of the type of interfacial systems analyzed. CHI (chitin) and TC molecules are embedded in-between CAL and HAP platelets, respectively. Such interfaces are then deformed under tensile and shear modes using established non-equilibrium molecular dynamics (NEMD) and steered molecular dynamics (SMD) schemes, with focus on measuring the interfacial shear strength in two separate deformation modes.

In order to analyze the effect of hydration on interface stress, water (WT) molecules are added to the interface region. The interface separation in Fig. 14.2 is chosen so as to have one layer of TC (or CHI) molecules in the interface region. Due to computational infeasibility of performing atomistic analysis of supercells with full-length TC (or CHI) molecules, only a segment of TC (or CHI) full-length molecule is used in the supercells. Stress–strain curve information is generated based on the well-known virial stress formulation using NEMD simulations. There are two loading directions (Fig. 14.3): direction along the molecule length (x-axis) and direction transverse to molecule length (y-axis). The simulated supercells are divided into slabs and three diagonal components of the pressure tensor in each slab are given in output, Fig. 14.3. The virial stress tensor of each slab and the overall system at the end of the equilibration is recorded as the stress tensor up to the point 20 % strain is achieved. The procedures make it possible to estimate how the measured stress of the loaded material system is distributed inside its interfacial regions and the stress between the interfaces can be obtained according to the behavior of the corresponding slab of the simulation system, Fig. 14.3.

NEMD simulations using the procedure is performed to predict interface strength by way of measuring stress–strain behavior of a thin block of atoms that are contained in the interfacial region. Such simulations cannot predict the effect of interface strength on interface separation mechanism. In order to understand such attribute, SMD simulations are performed. SMD simulations in the constant speed mode [7], were used to pull out the upper inorganic crystals (HAP or CAL) from the substrate inorganic crystals (HAP or CAL) in order to replicate the interfacial sliding process. Similar to the case of NEMD simulations, there are two loading directions: direction along the molecule length (x-axis) and direction transverse to molecule length (y-axis), Fig. 14.4. SMD force was applied to the center of mass of upper inorganic crystals in a chosen direction. The organic molecules (TC or CHI) and water molecules in the interface region were not under constraint. The substrate inorganic crystals were fixed on the bottom. In order to quantify the interface sliding process and failure in the interface region, a viscoplastic model [8] for interfacial sliding is introduced. The viscoplastic failure of the interfaces relates to the applied shear stress, τ, and to the shear velocity gradient (rate of shear deformation), $\frac{\partial V}{\partial d}$, after the yield stress, τ_0, is reached, as

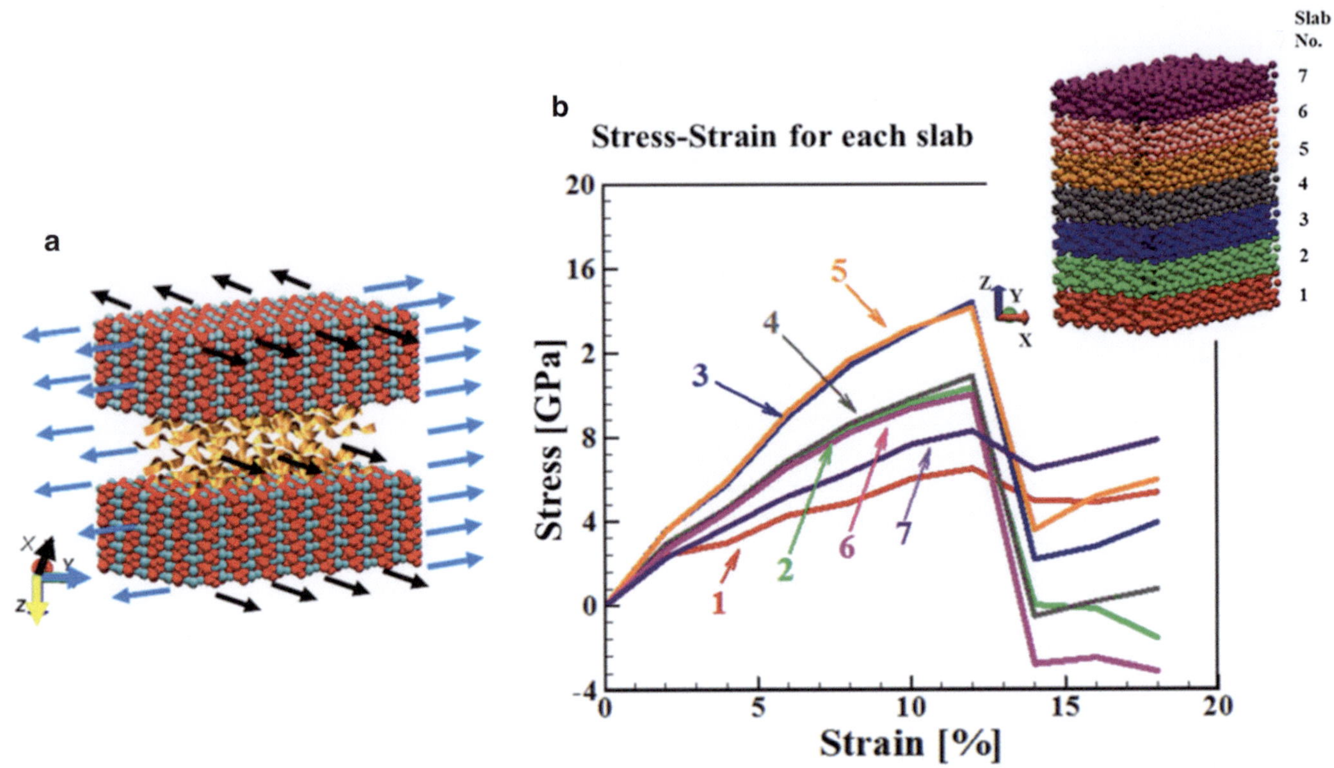

Fig. 14.3 (**a**) A schematic showing of loading condition of the interfacial material system and (**b**) stress–strain curve for each slab of the system from the *bottom* to the *top* layer

$$\tau = \tau_0 + \mu \frac{\partial \mathrm{V}}{\partial \mathrm{d}}. \tag{14.1}$$

Here, μ is the shear viscosity of interfacial sliding, and *d* is defined in Fig. 14.4.

In order to study the effect of shear rate on the viscous behavior of the interfacial systems, differently large force increments were only applied to the HAP-TC-HAP (and CAL-CHI-CAL) system with one layer of TC (and CHI) molecules, due to the limited space to perform the interfacial sliding with different shear rate. The shear viscosity of the HAP-HAP interface (Fig. 14.4) is calculated as 0.0232 Pa s. The viscosity of the slurries of HAP was reported as ~0.01–1.6 Pa s earlier by experiments performed [9], and the viscosity of montmorillonite hydrate was reported as ~0.008 Pa s by MD simulations [10]. The thickness of the interface region, *d*, can be determined from the plot. The yield shear stress, τY, is calculated from the critical force, *F0*, obtained by the curve fitting, and the interfacial area, *A*. The failure shear stress, τF, is calculated from the interface separation force divided by the interfacial area, *A*.

Shear stress plays an important role in the failure of the examined TC-HAP and CAL-CHI composite supercells. The lower and the upper bound of the shear strength of the interfacial material systems with organic interfaces or water interfaces could be defined by the yield shear stress, τY, and the failure stress, τF, which are used to characterize the plastic shear deformation, where $\tau_Y < \tau_F$. Figure 14.5a displays the yield shear stress (τY) calculated by means of SMD, the failure stress (τF) calculated using SMD, the interfacial shear strength of each of the TC-HAP interfacial material system shown in figure calculated using NEMD ($\sigma 1$), the shear strength of the organic TC phase region (or region of HAP cells with WT) ($\sigma 3$) calculated using NEMD, and the tensile mechanical strength of the organic phase region (or region of HAP cells with WT) ($\sigma 4$) calculated using NEMD, as functions of the interfacial components.

Figure 14.5a also compares effective viscosity and Young's moduli in dependence of the interfaces type. The HAP supercell consists of two HAP cells placed over each other separated by layer distance corresponding to PO_4 using in planes (7.5 Å). The WT supercell consists of the same HAP supercells, but with now water molecules separating the two. Figure 14.5b shows similar results in the case of CHI-CAL interfaces. In this case, the CAL supercell consists of two CAL cells placed over each other separated by layer distance corresponding to PO_4 using in planes (9.4 Å). The WT supercell consists of the same CAL supercells, but with now water molecules separating the two.

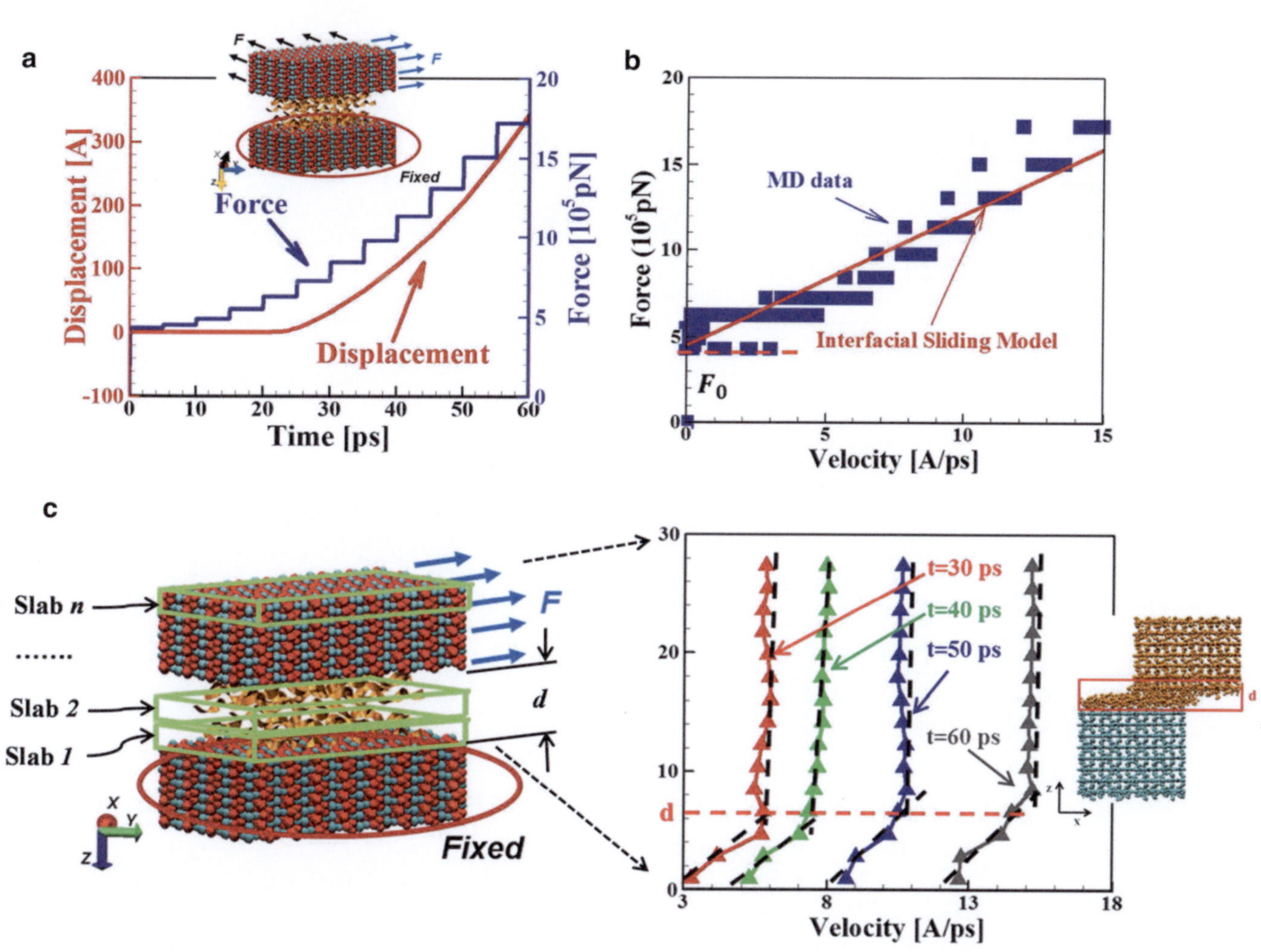

Fig. 14.4 (**a**) SMD force and displacement of the SMD structure as a function of time, with a schematic showing the loading condition of the SMD in constant force pulling mode, (**b**) curve fitting of the MD data with the viscoplastic model and (**c**) velocity profile at different time steps

Perhaps not surprisingly, the behavior of the two types of interfacial systems is quite similar to each other. The shear strength of organic phase (green line, $\sigma 3$) lies around the lower bounded line of the yield shear stress (blue dotted line, τY). The organic phases is the main contributor of the interfacial shear strength based on Fig. 14.5, where the interfacial shear strength (grey line, $\sigma 1$) matches closely with, sometimes a little higher than, the shear strength of the organic phases (green line, $\sigma 3$). However, it is always below the upper bound defined by the failure stress (red dotted line, τF) which indicates the "catastrophic failure" of the interfaces. The mechanical strength of the organic phases (yellow line, $\sigma 4$) usually lies between the lower and upper bounded line because shear deformation is usually the main contributor of the mechanical behavior of organic phases. Those points which are beyond the upper bound, concern hydrated organic interfaces (i.e., TC-WT or CHI-WT). This could be attributed to the much higher contribution of the shear interaction to the overall behavior as well as the higher shear viscosity of the interfacial material systems.

Whatever differing mechanisms are observable in the different investigated systems, all interfacial shear viscosities reported for the HAP-TC-HAP system are on the order of 10^{-2} Pa s. The viscosities obtained from our MD simulations are also much lower than experimental results of somehow related material systems, such as collagen gel with a viscosity of the order of 10^5 Pa s [11, 12]. The much higher shear rate in MD simulation leads to the measured lower viscosity values. Considered as the Newtonian fluid behavior, the shear rate ($\dot{\gamma}$) dependency of the viscosity (μ) of polymeric molecular structure is quite sensitive, i.e., increasing sharply as the shear rate decreases [13, 14]. Due to the computational capability of MD simulation, different shear rates varying from 10^7 to 10^9 1/s were performed on the HAP-TC-HAP system with one layer of TC molecules. Fitting with the widely used power law relation

$$\mu = \mathrm{B}\dot{\gamma}^{\mathrm{n}-1}, \tag{14.2}$$

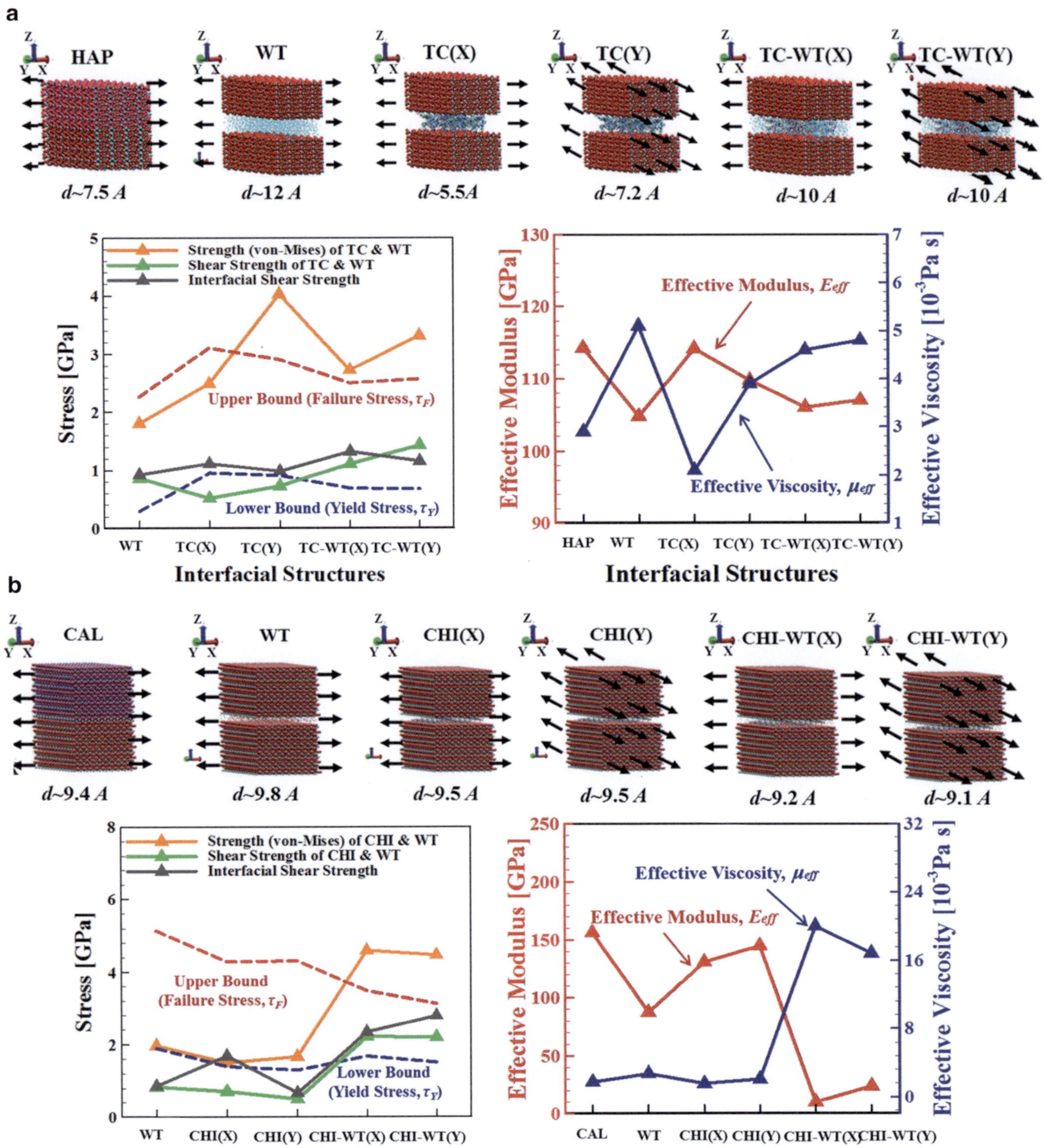

Fig. 14.5 Stress, Young's modulus, and effective viscosity, as functions of interfacial structures in the case of (**a**) TC-HAP interfaces and (**b**) CAL-CHI interfaces

with the parameters $B = 42.49$ Pa s and $n = 0.7034$, however, it is still not enough to capture the full picture of the viscosity-shear relationship because shear rate is still much higher than that used in creep or stress relaxation experiments (i.e., 10^{-6}–10^{-2} 1/s [13, 15], which is too low to be generated using MD simulation). The current study reports the viscous behavior of the bio interface systems at the infinite shear rate which can be used to estimate the properties beyond the observation range together with the material intrinsic property, i.e., zero-shear viscosity, μ_0, using the extrapolation methods, such as the cross model [13, 15],

$$\mu = \mu_\infty + \frac{\mu_0 - \mu_\infty}{1 + (C\dot{\gamma})^m} \tag{14.3}$$

where μ_∞ is the infinite shear viscosity obtained from the MD simulation, C is the cross time constant and m is known as the cross rate constant. With the viscosities of collagen materials at the lower shear rate from previous studies [11, 12, 16] and our infinity shear rate viscosity, the cross model parameters are obtained as $C = 3845.7$ s and $m = 2.002$. Figure 14.6 displays the viscosity versus shear rate behavior as a plot of $\log(\mu)$ versus $\log(\dot{\gamma})$. The cross model extrapolation captures the significant shear thinning behavior of the material within the low shear rate region and gives the idea of upscaling the MD viscosity results with the decreasing of the shear rate. The overall shear rate dependent viscous behavior of the material is predicted with the combination of the cross model and MD simulations.

Hard biological materials such as nacre, bone, and marine crustacean exoskeletons exhibit remarkable mechanical performance despite the fact that they are made up of relatively weaker constituents. In terms of the underlying mechanical principles for structural design of such materials, quite a few have been suggested. For example, one principle is the alignment of mineral-protein interfaces along the loading directions. MD study of TC-HAP biomaterials shows that a composite is best poised to handle the load if the protein molecules are in contact with mineral crystals having their longitudinal axis parallel to the mineral surface and along the loading direction of the composite. Second principle is the staggered arrangement of hard mineral crystals in soft protein matrix, leading to a unique mechanism of load transfer where crystals bear the normal load and protein transfers the load via shear. Third principle is that the failure of such polymer-ceramic type composites is dominantly peak strain dependent instead of peak strength. Also, presence of moisture at the interface enhances the stability and strength of such biomaterials by supporting the cross linking mechanism due to polar nature of water molecule.

One common feature which strongly stands out in most hard biological materials structures presence of interfaces at multiple levels of hierarchy. It seems that nature has designed these interfaces for optimum multifunctional performance during the course of evolution. Interfacial forces play a key role during deformation and failure of such biomaterials. Inorganic phases in the material systems carry the uniaxial tensile loading while the organic phases mainly carry the shear loading. Organic interfacial systems exhibit plastic shear deformation, the yield and failure shear stress define the lower and higher bound of the interfacial strength. Shear viscosity of the interfacial systems shows a highly shear rate dependent

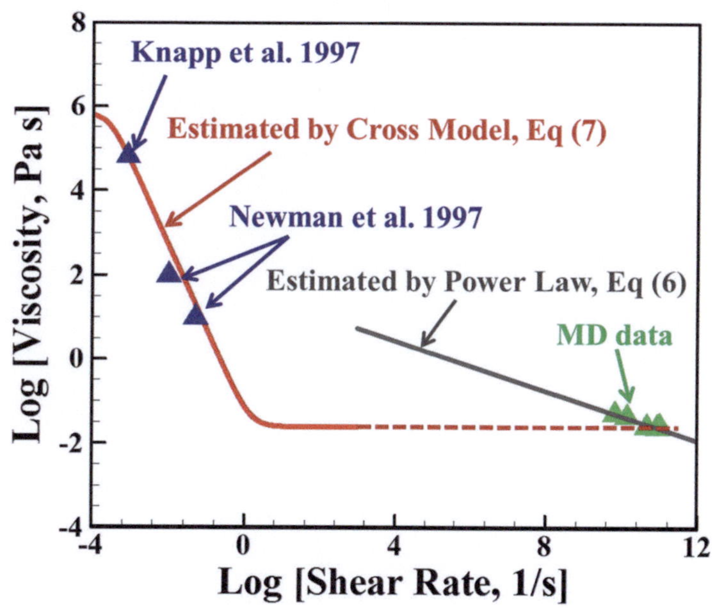

Fig. 14.6 Plot showing viscosity as a function of shear rate; references: Knapp et al. [11], Newman et al. [16]

behavior, however, the full picture of this behavior cannot be captured using MD simulation without the assistance from experimental technique and the extrapolation estimation. Interfacial interaction between the soft phase and hard phase is responsible for redistribution of stresses and directly affects the toughness and strength of the natural materials. Further, the design of the organic–inorganic interface along with the critical length of mineral constituent also contributes potentially in strengthening the biomaterials against failure and in affecting their overall mechanical performance.

References

1. Landis, W.J., Hodgens, K.J., Song, M.J., Arena, J., Kiyonaga, S., Marko, M., Owen, C., Mcewen, B.F.: Mineralization of collagen may occur on fibril surfaces: evidence from conventional and high-voltage electron microscopy and three-dimensional imaging. J. Struct. Biol. **117**, 24–35 (1996)
2. Landis, W.J., Hodgens, K.J., Arena, J., Song, M.J., McEwen, B.F.: Structural relations between collagen and mineral in bone as determined by high voltage electron microscopic tomography. Microsc. Res. Tech. **33**(2), 192–202 (1996)
3. Fratzl, P., Fratzlzelman, N., Klaushofer, K., Vogl, G., Koller, K.: Nucleation and growth of mineral crystals in bone studied by small-angle X-ray scattering. Calcif. Tissue Int. **48**(6), 407–413 (1991)
4. Weiner, S., Talmon, Y., Traub, W.: Electron diffraction of mollusc shell organic matrices and their relationship to the mineral phase. Int. J. Biol. Macromol. **5**(6), 325–328 (1983)
5. Al-Sawalmih, A., Li, C., Siegel, S., Fabritius, H., Yi, S., Raabe, D., Fratzl, P., Paris, O.: Microtexture and chitin/calcite orientation relationship in the mineralized exoskeleton of the American lobster. Adv. Funct. Mater. **18**(20), 3307–3314 (2008)
6. Qu, T., Tomar, V.: Nanomechanics based investigation into interface -thermomechanics of collagen and chitin based biomaterials. In: Proceedings of the Society of Engineering Science 51st Annual Technical Meeting. Purdue University Libraries Scholarly Publishing Services, West Lafayette, 1–3 Oct 2014
7. Phillips, J.C., Braun, R., Wang, W., Gumbart, J., Tajkhorshid, E., Villa, E., Chipot, C., Skeel, R.D.: Scalable molecular dynamics with NAMD. J. Comput. Chem. **26**, 1781–1802 (2005)
8. Frankland, S., Harik, V.: Analysis of carbon nanotube pull-out from a polymer matrix. Surf. Sci. **525**(1), L103–L108 (2003)
9. Lelievre, F., Bernache-Assollant, D., Chartier, T.: Influence of powder characteristics on the rheological behaviour of hydroxyapatite slurries. J. Mater. Sci. Mater. Med. **7**(8), 489–494 (1996)
10. Ichikawa, Y., Kawamura, K., Fujii, N., Nattavut, T.: Molecular dynamics and multiscale homogenization analysis of seepage/diffusion problem in bentonite clay. Int. J. Numer. Methods Eng. **54**(12), 1717–1749 (2002)
11. Knapp, D.M., Barocas, V.H., Moon, A.G., Yoo, K., Petzold, L.R., Tranquillo, R.T.: Rheology of reconstituted type I collagen gel in confined compression. J. Rheol. **41**(5), 971–993 (1997) (1978-present)
12. Barocas, V.H., Moon, A.G., Tranquillo, R.T.: The fibroblast-populated collagen microsphere assay of cell traction force—Part 2: Measurement of the cell traction parameter. J. Biomech. Eng. **117**(2), 161–170 (1995)
13. Dealy, J.M., Wang, J.: Melt rheology and its applications in the plastics industry. Springer, Netherlands (2013)
14. Bylund, G., Pak, T.: Dairy processing handbook. Tetra Pak Processing Systems AB, Lund (2003)
15. Franck, A.: Understanding Rheology of Thermoplastic Polymers. TA Instruments 2004
16. Newman, S., Cloitre, M., Allain, C., Forgacs, G., Beysens, D.: Viscosity and elasticity during collagen assembly in vitro: relevance to matrix-driven translocation. Biopolymers **41**(3), 337–347 (1997)

Chapter 15
The Growth and Mechanical Properties of Abalone Nacre Mesolayer

Yan Chen, MariAnne Sullivan, and Barton C. Prorok

Abstract Abalone nacre has long been interesting to material engineers due to its fascinating structure and outstanding mechanical behavior. It is a self-organized biogenic mineral composite which has a hierarchical architecture composite. The main component of this structure is "block and mortar" like tablets. Between the tablets structures is a layered structure which consist of both organic and inorganic parts called mesolayer. This research focus on the growth behavior of the mesolayer structure and its mechanical influence on nacre. Fresh-grown nacre is harvested by the "flat pearl" technique under different temperature schemes. SEM observation of those nacre reveals temperature scheme will inhibits the growth of tablets structure, and form other layered structure. The mechanical property of harvested "flat pearl" is tested by nano-indentation. In this work, we aim to show the mechanic significance of mesolayer in nacre structure, and hope to establish its strengthening and growth mechanism.

Keywords Biomimetic • Nanoindentation • Biomineralization • Multilayer composites

15.1 Introduction

Nacre is a structure that is found in many animals including mollusk or bivalves. It is also called mother-of-pearl. One can find it in the underside of shells of such animals, attached to the calcite shells. From ancient time, they were used as ornaments, jewelries because of its iridescent appearance. However, what makes it unique to the world as materials is its strength. The mechanical behavior of nacre is different for different types of nacres. The most extraordinary one is the nacre of abalone, a gastropod mollusk [1]. This kind of nacre is generally called "column nacre" due to the columnar appearance of its cross section [1, 2].

This nacre is composed by 95 wt % of calcite (aragonite polymorph), and 5 wt % of protein. They appear as a layered structure—a layer of aragonite "tiles" separated by the organic thin layer [1]. The structure of wild abalone nacre has a hierarchical complex structure. Between the nacreous aragonite layers, there are several layers of non-nacre structure which is called mesolayer. This structure is consist of one layer of single crystal aragonite and one layer organic layer [3].

The mechanical behavior of nacre is tested by numerous studies. Their results are very different because of different specimen conditions (wet or dry, shell with nacre or nacre only etc.), but all show a significant improvement compared to the aragonite [1, 4]. However, there are very little researches reveal the importance of the mesolayer structure on the mechanical behavior of nacre structure. This may be due to the fact that many research conducted on the farm raised nacre while a farm raised nacre do not have such structure [5]. The difference between those two is very obvious in the cross sectional SEM images as shown in Fig. 15.1.

In the previous studies in our group, we have already shown that the elastic modulus tested by nanoindentation of wild and farm raised nacre are significantly different. The farm raised one is almost 20 GPa lower than the wild one [5]. One should notice that although the nanoindentation test result cannot be directly interpreted as the elastic modulus of one material [6], higher elastic modulus of wild abalone does show its superior mechanical behavior.

Y. Chen (✉) • M. Sullivan • B.C. Prorok, Ph.D.
Department of Mechanical Engineering, Auburn University, 275 Wilmore Laboratories, Auburn, AL 36849, USA
e-mail: YZC0045@auburn.edu

S.A. Tekalur et al. (eds.), *Mechanics of Biological Systems and Materials, Volume 6*,
Conference Proceedings of the Society for Experimental Mechanics Series, DOI 10.1007/978-3-319-21455-9_15

Fig. 15.1 Farm raised abalone nacre (*right*), compare with wild abalone nacre (*left*) with the mesolayers

The abalone nacre, is not just one type of composite material, it is also two types of composite material in different scale. That means, on nano scale, there is aragonite–protein composite, and on micron scale, there is mesolayer–nacre layer structure. Understanding the formation and the mechanic significant of this structure can be applied on many areas of material engineering: design of protective layer, mineralization of single crystal, polymer–ceramic composite etc.

15.2 Experimental Procedure

Abalones were shipped from abalone farm in California and feed in aquariums. Our Lab has ten different aquariums and the temperature of them are controlled individually by chillers, as shown in Fig. 15.2. For a more precise control of temperature, chillers are set in Fahrenheit scale.

During the shipment, abalones were kept in a low temperature, then, they were revived in 55 °F. We were successfully keep several abalones under such lab conditions.

The method to study the growth behavior of abalone nacre is the "flat pearl" technic pioneered by researchers in UC Santa Barbara [7]. To obtain the growth behavior of nacre, a glass slide is attached under the mantle of the animal by underwater glue, as shown in Fig. 15.3. After some time (about a week), abalone will cover the glass slide with its mantle and the growth of nacreous structure will appear on the glass slide. Fractured cross section of the glass slide was observed under SEM.

Our pervious study has already shown that the change of temperature will induce a structure change during the growth of nacre. However, the previously induced structure is not a complete mesolayer, but only a part of it. A further discussion of our pervious study will be present latter.

The mechanical behavior of wild, farm and lab raised abalone nacre is studied by Nanoindentation on MTS Nanoindenter XP with continuous measurements. The tested surface is perpendicular to the growth direction of nacre.

15.3 Results and Discussion

Figure 15.4 shows our nanoindentation data on wild and farm raised nacre. The elastic modulus tested on wild nacre is comparable with the elastic modulus of aragonite crystal [4]. Consider the fact that nacre is 95 wt % of aragonite, this result shows that under indentation of 1000 nm displacement, the structure of nacre is still intact under the indent.

The lower elastic modulus of a farm raised nacre can be explained by its relatively homogeneous structure. Lack of mesolayer can makes it more susceptive to cracking.

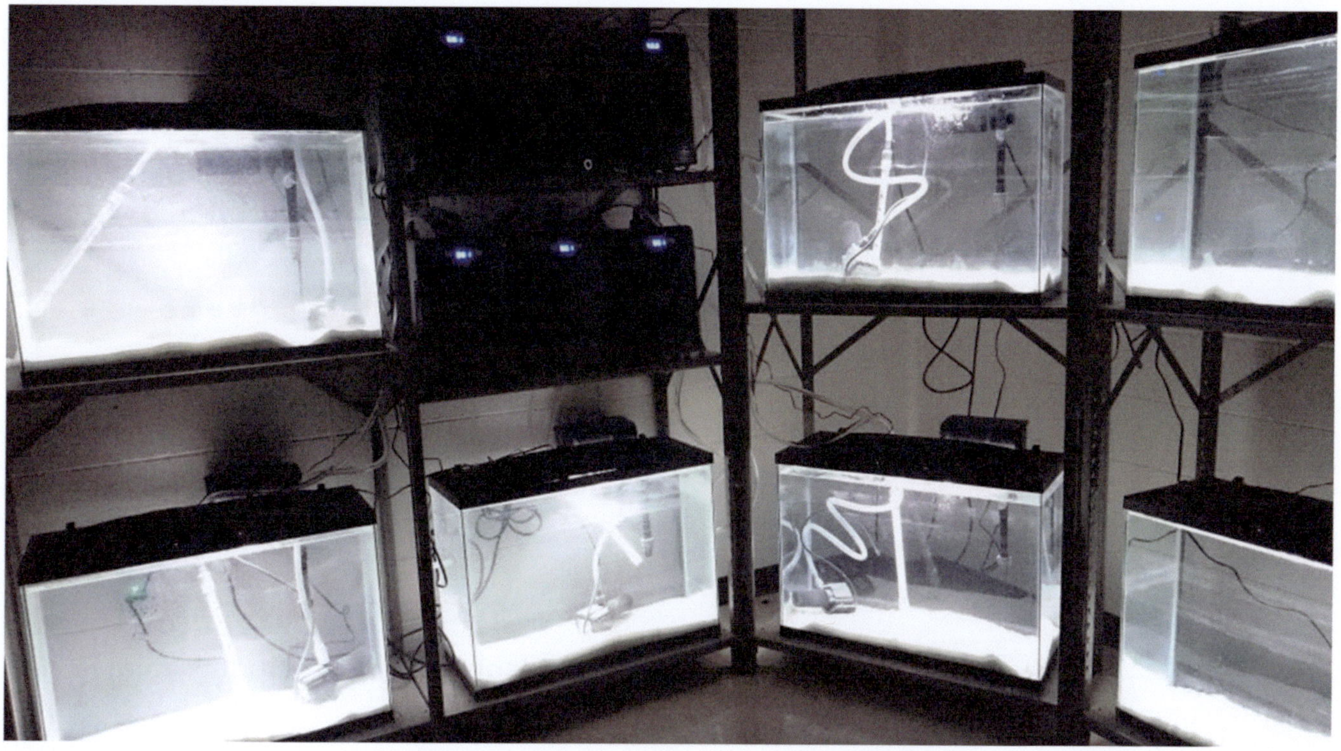

Fig. 15.2 Aquarium for abalone

Fig. 15.3 Attached glass slide on the growth front of nacre

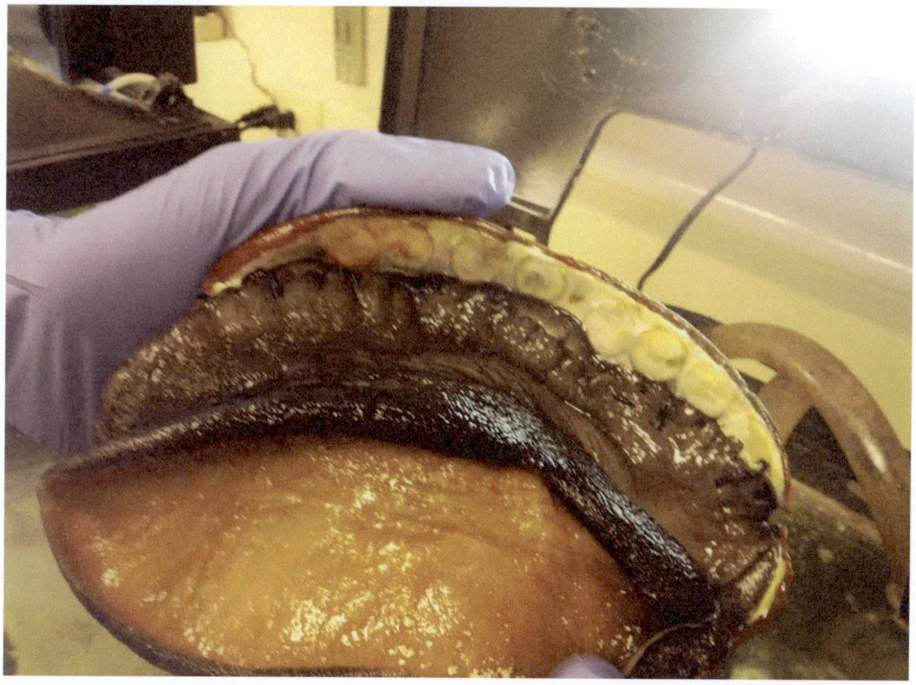

Figure 15.5 shows the cross sectional indentation data on the wet nacre and organic part of mesolayer. It is obvious that compared to nacre, the mesolayer organic part is extremely compliance. This property can be interpreted as the function of mesolayer releasing the strain inside the nacre to prevent crack forming.

The structure formed by the nacre can be influenced by the temperature during forming as we have shown in our previous studies. Figure 15.6 is a backscatter image of a wild nacre and the nacre we bio-fabricated in our lab. Temperature schedule of its forming is shown in the previous study [5].

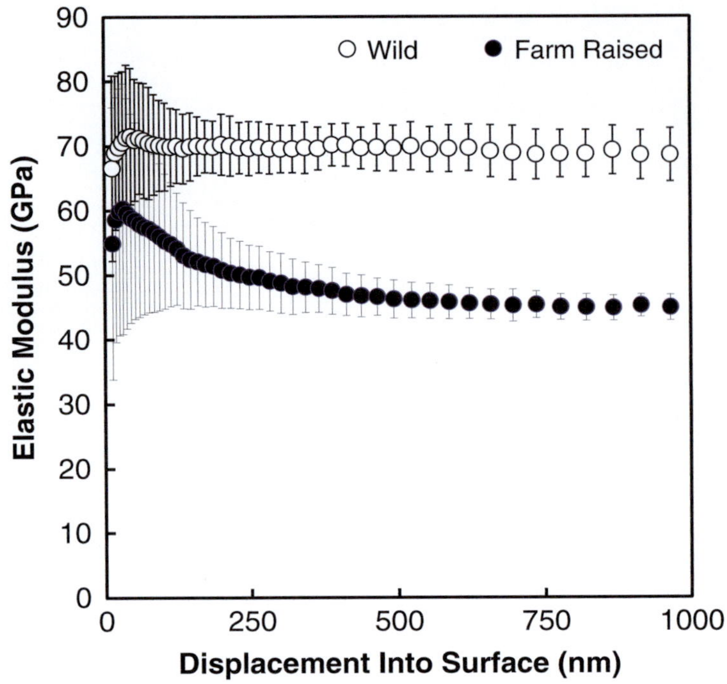

Fig. 15.4 Nanoindentation of elastic modulus on surface of wild and farm raised abalone

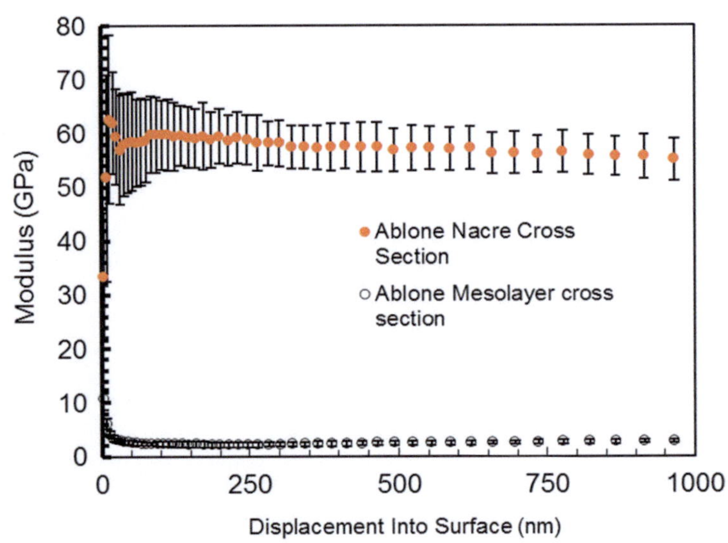

Fig. 15.5 Nanoindentation of elastic modulus on cross section of wild nacre

Although we have already shown that in our lab we can form a mesolayer structure, in backscatter image in Fig. 15.6, we can see that previous study is only forming the block like layer of the mesolayer.

We believe that the forming of the complete mesolayer structure can be accomplished by a larger scale of experiment with our recent improvement of the lab environment.

15.4 Conclusion

It is concluded that abalone can be kept in a lab environment, and the nacre structure can be changed by the changing of temperature. The mesolayer of nacre is not just structurally different with the nacreous structure, but has a significant different mechanical behavior.

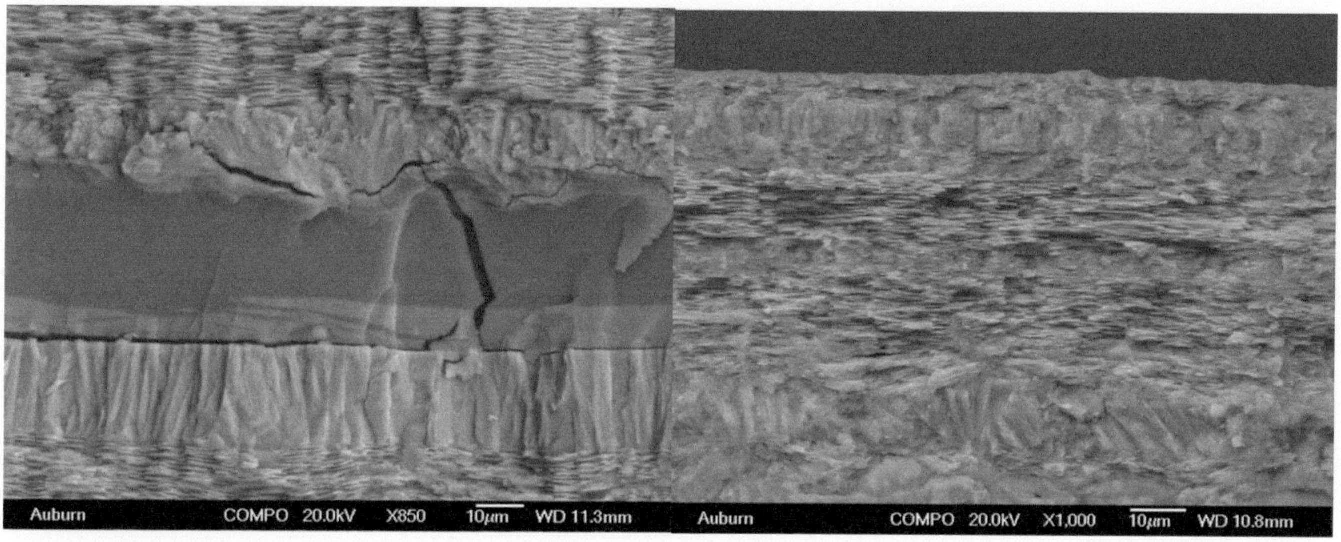

Fig. 15.6 Backscatter Phase contrast of wild nacre (*left*), Lab formed nacre (*right*)

Although the nacre structure is always perceived in nano scale, we here presented that it can also be seen as a composite material in micro scale: the "gradually failing" nacreous layer can be seen as metal like material, and low elastic modulus ceramic block like structure, with a low elastic modulus organic layer in between. This insight can provide us with some important design principals of any protective material with architecture structure.

References

1. Zhang, X., Fan, Z., Lu, Q., Huang, Y., Kaplan, D.L., Zhu, H.: Hierarchical biomineralization of calcium carbonate regulated by silk microspheres. Acta Biomater. **9**, 6974–6980 (2013)
2. Barthelat, C.-M., Li, C.C., Espinosa, H.D.: Mechanical properties of nacre constituents and their impact on mechanical performance. J. Mater. Res. **21**, 1977–1986 (2006)
3. Zaremba, C.M., et al.: Critical transitions in the biofabrication of abalone shells and flat pearls. Chem. Mater. **8**, 679–690 (1996)
4. Zhang, J., Zhang, Y., Xu, K., Ji, V.: Young's modulus surface and Poisson's ratio curve for orthorhombic crystals. J. Chem. Crystallogr. **38**(10), 733–742 (2008)
5. Sullivan, M.A., Prorok, B.C.: Controlling abalone shell architecture with temperature. In: Mechanics of Biological Systems and Materials, vol. 7, pp. 27–31. Springer, Heidelberg (2015)
6. Zhou, B., Prorok, B.C.: A new paradigm in thin film indentation. J. Mater. Res. **25**(09), 1671–1678 (2010)
7. Fritz, M., Belcher, A.M., Radmacher, M., Walters, D.A., Hansma, P.K., Stucky, G.D., Morse, D.E., Mann, S.: Flat pearls from biofabrication of organized composites on inorganic substrates. Nature **371**(6492), 49–51 (1994)

Chapter 16
Novel Bending Fatigue Testing of Small Medical Device Cables

Haitao Zhang and Bernard Li

Abstract Neuromodulation is among one of the fastest-growing areas of the medical device industry. Neuro stimulators are being used to provide therapies to patients with chronic pain, movement disorders, overactive bladders, and psychiatric disorder symptoms. In these therapies, well-controlled stimulation current is delivered to the target nerves through implantable leads. The implantable leads must have sufficient fatigue durability to last for many years of implantation.

The implantable leads are composed of insulation polymer jackets and conductor coils/cables. Conductor cables are generally very small and made from MP35N alloys (Co–35Ni–20Cr–10Mo). These conductor cables are more susceptible to fatigue failure than the insulation polymer jackets.

A novel bending fatigue testing method was developed and used to test small medical device cables. 1×7 standard MP35N and low Ti MP35N cables are tested. The generalized S–N curve, in the form of curvature amplitude as a function of cycle to failure was obtained. The results show that low Ti MP35N cable has better fatigue performance than the standard MP35N cable. A SEM image of fractured surface shows that cable fatigue is dominated by crack initiation. The crack initiation is induced by shear deformation. This is a typical crack initiation mechanism under high plasticity deformation.

Keywords Fatigue • Cable • Medical device • Testing • Neuromodulation

16.1 Introduction

Neuromodulation is among one of the fastest-growing areas of the medical device industry. Neuro stimulators are being used to provide therapies to patients with chronic pain [1–3], movement disorders [4–6], overactive bladders [7–9] and psychiatric disorder symptoms. In these therapies, well-controlled stimulation current is delivered to the target nerves through implantable leads. The implantable leads must have sufficient fatigue durability to last for many years of implantation.

The implantable leads are composed of insulation polymer jackets and conductor coils/cables. For example, helically-wound cobalt-based alloy, MP35N (Co–35Ni–20Cr–10Mo), cable conductors are widely used in spinal cord stimulation lead design. These conductor cables are more susceptible to fatigue failure than the insulation polymer jackets. Figure 16.1 shows one example where MP35N cables were used as lead conductors for spinal cord stimulation.

Fatigue testing methods used to obtain a material S–N curve typically involve using large coupon samples which have a characteristic size around several millimeters [10]. It has been shown that fatigue behavior of small medical wires can be significantly different from that of their base materials [11]. Some efforts have been made to characterize the fatigue of conductor cables with a rotary bending method [12] and a flex bending method [13] by adapting ASTM E796-94 (Reapproved 2000) [14]. One limitation of the rotary bending method is that the loading condition needs to be in the elastic range. Some tracking fixtures [15] have been added to the rotary bending method so the method can apply to samples with some plastic deformation. However the additional fixture may induce no clinically-relevant boundary constraint for lead cables. The flex method adapted from ASTM E796-94 is able to test samples with plastic deformation. However the mandrels induce some additional boundary constraints on the cable wire, which might lead to less conservative fatigue life compared to rotating bending testing as shown in [13].

H. Zhang (✉) • B. Li
Medtronic, 7000 Central Ave. NE, Minneapolis, MN 55432, USA
e-mail: haitao.zhang@medtronic.com

S.A. Tekalur et al. (eds.), *Mechanics of Biological Systems and Materials, Volume 6*,
Conference Proceedings of the Society for Experimental Mechanics Series, DOI 10.1007/978-3-319-21455-9_16

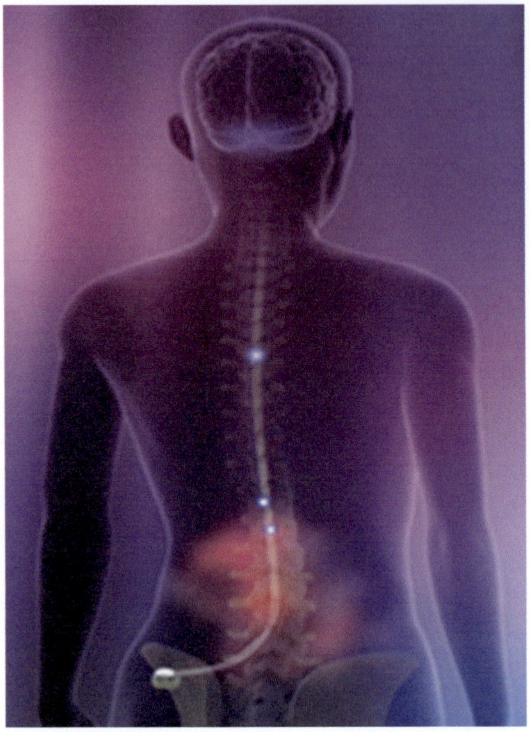

Fig. 16.1 Example applications of MP35N cables in a medical device

In this paper, a novel bending fatigue testing method is developed to characterize the fatigue behavior of small MP35N cables. This method also incorporates a newly-developed Labview-based conductor fracture-monitoring system to monitor and detect cable fracture automatically. The fatigue result for MP35N cables is analyzed with a generalized S–N curve. A scanning electron microscope (SEM) was used to analyze the fatigue fracture surface of the samples.

16.2 Experiment

16.2.1 Materials

The schematic description of the cable architecture for 1×7 MP35N and 1×7 low Ti MP35N cables used in this study is shown in Fig. 16.2. The 1×7 cable consists of seven wires twisted in a 1×7 configuration and coated with ETFE as an electrical insulator layer. The 1×7 cable has an overall diameter of 0.005″. Each of the seven wires has a diameter of 0.00167″. The thickness of the ETFE coating is 0.001″.

16.2.2 Fatigue Testing

A new testing system was developed to test MP35N 1×7 and low Ti MP35N cables. Figure 16.3 shows the system design of the test system. The system consists of a Bose Electro Force 3200 Series high-speed linear fatigue testing machine, a computer controlling the fatigue machine, customized electric monitoring circuitry and monitoring computer, and a customized fixture (shown in Fig. 16.4, with a cable sample).

Each cable has a length of 11 in. On both ends, about 0.5 in. of ETFE coating was stripped off with tweezers to expose cable metal wires, which will be connected to the resistance monitoring circuit before the testing starts. The sample is attached to the fixture with a piece of tape. The gage length of the sample is 1.57 in.

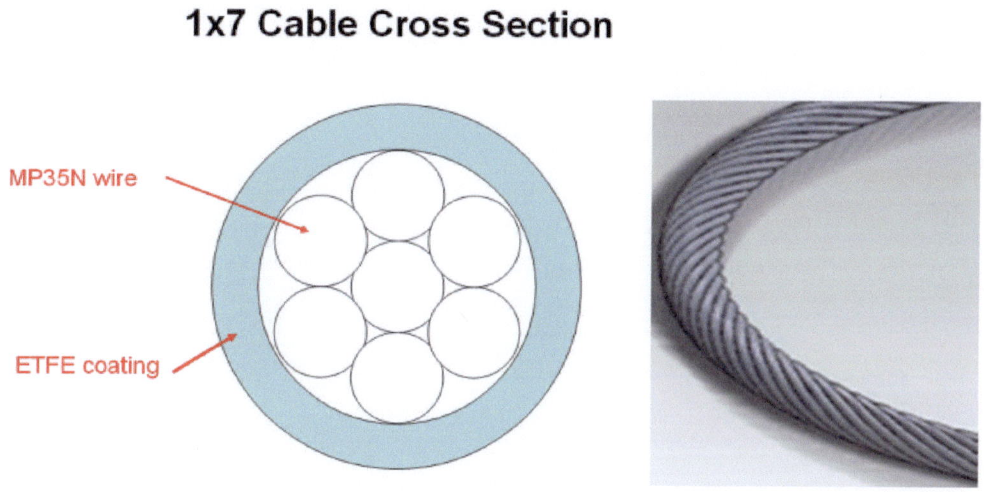

Fig. 16.2 Schematic illustration of 1×7 cable

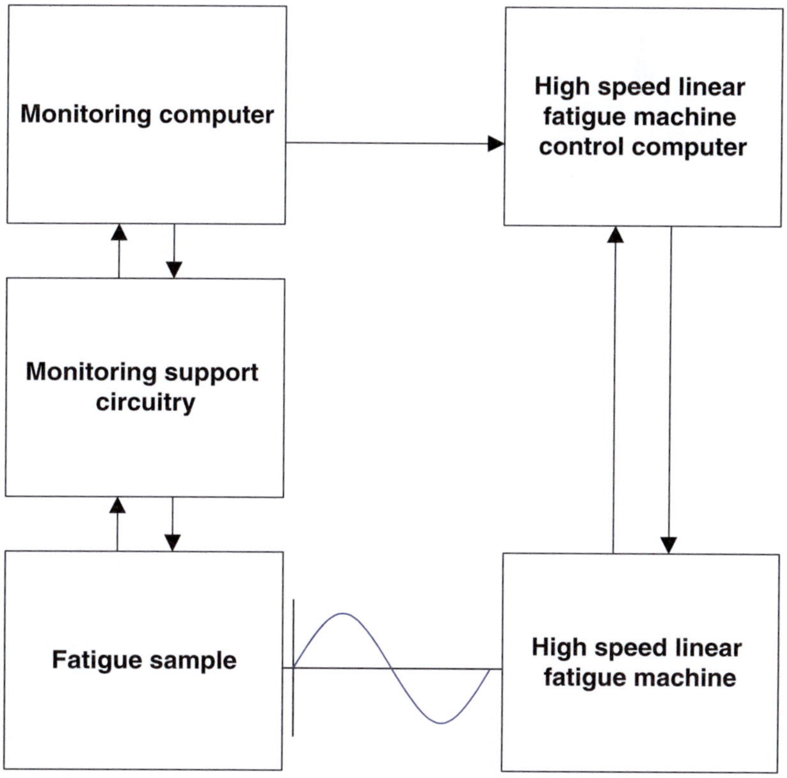

Fig. 16.3 System diagram of cable bending fatigue test system

As shown in Fig. 16.5, the fixture was then mounted to the Bose Electro Force 3200 machine. The initial gap between the top fixture and lower fixture was set to 0.25 in. A gage block was attached next to the sample to serve as a reference for data reduction later, and a digital image (I0) was then taken. The sample was then bent from the initial radius (R_1) to a pre-defined tighter radius (R_2), and a second image (I1) was then taken. After that, the gap was moved back to 0.25 in. The sample was subsequently attached to the circuit monitoring system and fatigue testing was started at a frequency of 30 Hz. The testing stopped when the cable fractured. The fatigue fracture surface of the test sample was then analyzed in an SEM, JEOL 6480 L, operated at 5 kV.

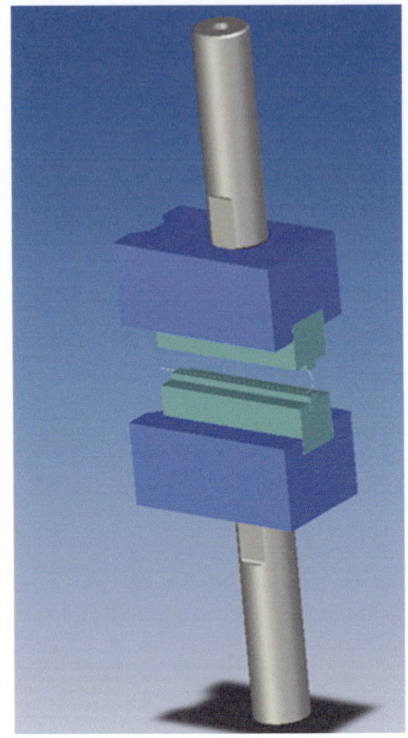

Fig. 16.4 Schematic description of test fixture with cable sample

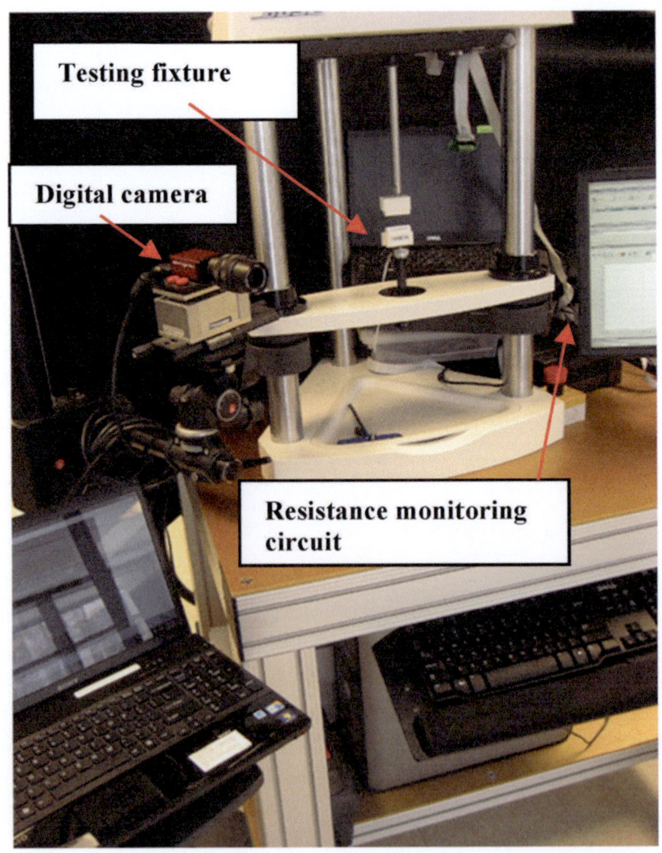

Fig. 16.5 Cable bending fatigue test system

16.2.3 *Data Analysis*

Figure 16.6 shows the cable deformation during the fatigue. The R_1 and R_2 are measured from image I0 and I1 respectively. The images are first imported into Solidworks software. The radiuses, R_1 and R_2 to are then measured with the gage block as a calibration reference. The smart dimension function of the Solidworks software is used.

The curvature amplitude is defined as

$$\Delta C = \left(\frac{1}{R_2} - \frac{1}{R_1}\right)/2 \tag{16.1}$$

The equivalent "strain" life is expressed as ΔC as a function of cycle to failure.

16.3 Results

16.3.1 *Tensile Properties*

MP35N is a Co-based alloy. It has base composition Co–35Ni-20Cr–10Mo. The standard MP35N alloy has less than 1 % Ti content and the low Ti MP35N alloy has a lower content of Ti, less than 0.01 % Ti. The low Ti MP35N has a better fatigue performance than MP35N and is widely used in the medical device industry.

The tensile property of the MP35N cable wire was tested in a single wire that was processed under the same conditions that were used for the cable. The tensile test was performed on a 10 in long gage length and with speed of 0.5 in/minute.

The wire properties listed in Table 16.1 are the result of averaging results from six tests.

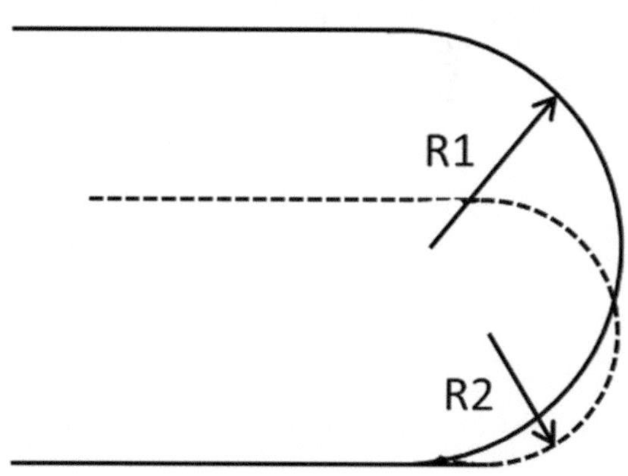

Fig. 16.6 Cable deformations during fatigue testing

Table 16.1 Tensile properties of MN35N and low Ti MP35N 1×7 cable

Material	UTS (ksi)	TYS (ksi)	Elongation (%)	RA (%)	E (Msi)	Break load (lb)
MP35N 1×7 cable	327	320	1.2	17	34	4.8
Low Ti MP35N 1×7 cable	239	210	1.8			3.5

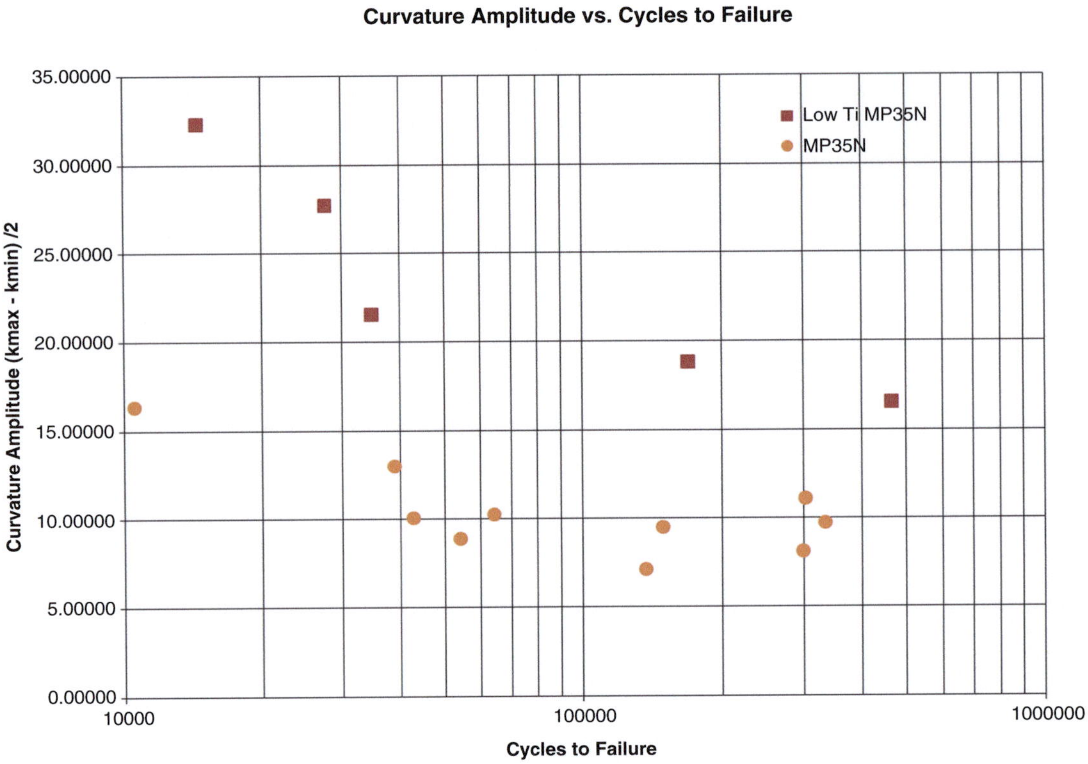

Fig. 16.7 Generalized bending fatigue S–N curves for 1×7 MP35N and low Ti MP35N cables

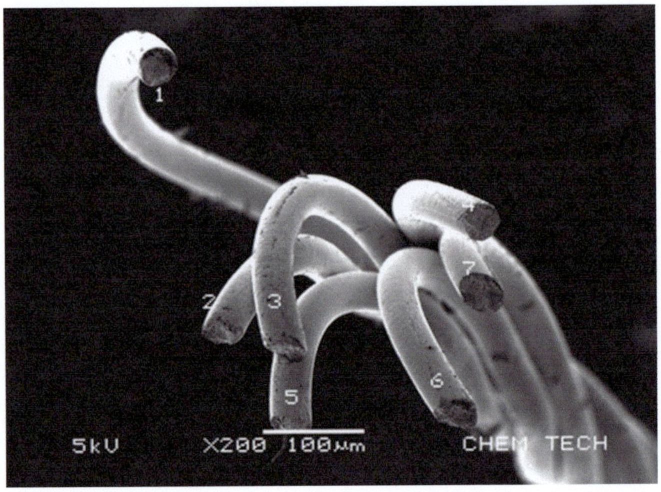

Fig. 16.8 SEM image of fractured surface at low magnification with fatigue cycle of 14,481

16.3.2 Fatigue

The generalized S–N curves for MP35N 1×7 cable and Low Ti MP35N cables are plotted in Fig. 16.7, in the form of curvature amplitude as a function of cycle to failure. As expected, higher curvature amplitude, which corresponds to a sharper bend, reduces cable fatigue life. Low Ti MP35N cable has better fatigue performance than the standard MP35N cable.

Fracture surfaces of fatigue-fractured samples were examined in an SEM. Figures 16.8, 16.9, and 16.10 show the cross section of fatigue-fractured low Ti MP35N 1×7 cable with cycle number 14,481. In Fig. 16.8, the image shows the wire is plastically deformed during U-bend testing, which indicates the fatigue stress was in the plastic region. In Fig. 16.9, the image of the wire shows a two-sided fatigue initiation with the fatigue initiated from the 1 o'clock and 7 o'clock positions.

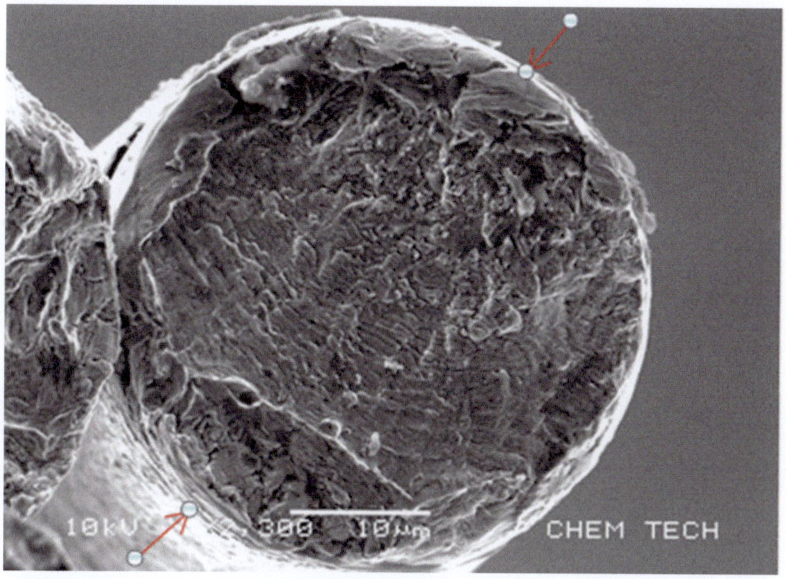

Fig. 16.9 SEM image of fractured surface of a single wire with fatigue cycle of 14,481

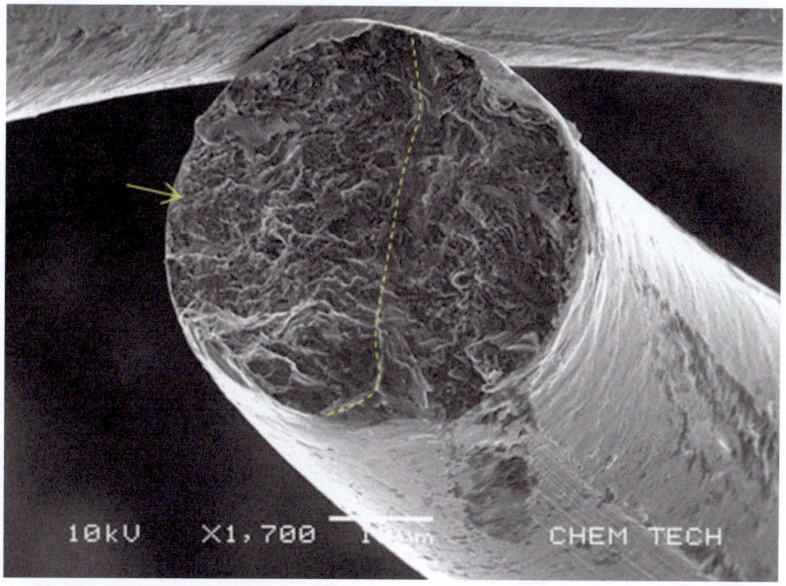

Fig. 16.10 SEM image of fractured surface of a single wire with fatigue cycle of 14,481

The image shows visible fatigue striations from 1 o'clock propagated to the 7 o'clock position. The form of the striation indicates that the deformation is in the plastic range of the wire. The image shows that cable fatigue is dominated by crack initiation and micro cracks. The crack initiation mechanism is plastic deformation induced by the high shear stresses. This is a typical crack initiation mechanism under high plasticity deformation.

The SEM observation shows that each of the seven wires has different fatigue features which are caused by the stresses on the wires at different fatigue stages due to the geometric configuration of the cable. Figure 16.10 shows another wire fatigue fracture image. The crack was initiated from both the left and right edges. They propagated toward the middle and met at the yellow dashed line. There are no foreign particles or other defects observed at the crack initiation site. This indicates that the crack initiation is due to plastic deformation of the wire. The plastic deformation can be seen near the crack surface on the wire.

The cable with a higher fatigue cycle number shows a slightly different fatigue feature. Figures 16.11a, b are the fatigue fracture images from a cable with fatigue cycles of 465,620. The fracture surface shows plastic deformation as shown in Fig. 16.11a. In Fig. 16.11b, which shows two sides fatigue initiation, the crack propagation was not observed. The dominant fatigue failure is the micro-cracking initiation and propagation. Some areas show fine fatigue striation propagated to different directions.

Fig. 16.11 SEM image of fractured surface of a single wire with cycle 465,620

16.4 Conclusion

A novel bending fatigue testing method was developed and used to test small medical device cables. 1×7 standard MP35N and Low Ti MP35N cables were tested. The generalized S–N curve, in the form of curvature amplitude as a function of cycle to failure, was obtained. The results show that low Ti MP35N cable has better fatigue performance than the standard MP35N cable. A SEM image of the fractured surface shows that cable fatigue is dominated by crack initiation. The crack initiation is induced by shear deformation. This is a typical crack initiation mechanism under high plasticity deformation.

References

1. Barolat, G., Sharan, A.D.: Spinal cord stimulation for chronic pain management. Semin. Neurosurg. **15**, 151–175 (2004)
2. Cameron, T.: Safety and efficacy of spinal cord stimulation for the treatment of chronic pain: a 20-year literature review. J. Neurosurg. Spine **100**, 254–267 (2004)
3. Manchikanti, A., Boswell, M., Datta, S., Fellows, B., Abdi, S., Vijay, V., Benyamin, R., Falco, F., Helm, S., Hayek, S., Smith, H.: Comprehensive review of therapeutic interventions in managing chronic spinal pain. Pain Physician **12**, E123–E198 (2009)
4. Blomstedt, P., Hariz, M.I.: Deep brain stimulation for movement disorders before DBS for movement disorders. Parkinsonism Relat. Disord. **16**, 429–433 (2010)
5. Benabid, A.L., Chabardes, S., Torres, N., Piallat, B., Krack, P., Fraix, V., Pollak, P.: Review functional neurosurgery for movement disorders: a historical perspective. Prog. Brain Res. **175**, 379–391 (2009)
6. Pizzolato, G., Mandat, T.: Deep brain stimulation for movement disorders. Front. Integr. Neurosci. **6**, 2–5 (2012)
7. Kohli, N., Patterson, D.: InterStim® therapy: a contemporary approach to overactive bladder. Rev. Obstet. Gynecol. **2**, 18–27 (2009)
8. Brazzelli, M., Murray, A., Fraser, C.: Efficacy and safety of sacral nerve stimulation for urinary urge incontinence: a systematic review. J. Urol. **175**, 835–841 (2006)
9. Sherman, N., Jamison, M., Webster, G., Amundsen, C.L.: Sacral neuromodulation for the treatment of refractory urinary urge incontinence after stress incontinence surgery. Am. J. Obstet. Gynecol. **192**, 1643–1648 (2005)
10. Akiniwa, Y., Stanzl-Tschegg, S., Mayer, H., Wakita, M., Tanaka, K.: Fatigue strength of spring steel under axial and torsional loading in the very high cycle regime. Int. J. Fatigue **30**, 2057–2063 (2008)
11. Li, B., Zhang, H., Wang, L.: MPMD 2011. http://asm.confex.com/asm/mpmd11/webprogram/Paper28326.html
12. Altman, P.A., Meagher, J.M., Walsh, D.W., Hoffmann, D.A.: Rotary bending fatigue of coils and wires used in cardiac lead design. J. Biomed. Mater. Res. **43**, 21–37 (1998)
13. Lewandowski, J., Varadarajan, R., Smith, B., Tuma, C., Shazly, M., Vatamanu, L.: Tension and fatigue behavior of 316LVM 1×7 multi-strand cables used as implantable electrodes. Mater. Sci. Eng. A Struct. Mater. **486**, 447–454 (2008)
14. ASTM E 796-94 (Reapproved 2000). Standard test method for ductility testing of metallic foil. Annual Book of ASTM Standards, 1994.
15. Scheiner, A., Thomas, J., Kicher, T.: A study of the fatigue properties of small diameter wires used in intramuscular electrodes. J. Biomed. Mater. Res. **25**, 589–608 (1991)

Chapter 17
Complex Shear Modulus of Hydrogels Using a Dynamic Nanoindentation Method

Riaz Akhtar, Emily R. Draper, Dave J. Adams, and Holger Pfaff

Abstract The micromechanical properties of soft tissues and materials are of considerable interest for biomedical applications. Nanoindentation is a powerful technique for determining localized material properties of biological tissues and has been used widely for hard tissue and material characterization. However, the technique is much more challenging when utilized for soft tissues due to their compliance as well as due to the limitations of commercial instruments which were originally developed for stiff, engineering materials. This study explores the use of a dynamic indentation method with a cylindrical punch (100 μm diameter) to characterize gelatin gel and low molecular weight hydrogels. A Keysight Technologies DCM II actuator is used with the Continuous Stiffness Measurement (CSM) to determine the complex shear modulus of these gels. The method overcomes surface detection issues with standard quasi-static nanoindentation as a change in phase angle can be used to accurately detect the sample surface. The data collected in this study are found to be comparable with macroscopic rheology and demonstrates the utility of the method for characterization of hydrogels.

Keywords Nanoindentation • Hydrogels • Shear modulus • Dynamic indentation

17.1 Introduction

Nanoindentation is being used increasingly to characterise the mechanical properties of soft materials such as hydrogels [1] and compliant polymers [2] as well as soft tissues [3, 4]. For biomedical applications, accurate characterisation of the mechanical properties of hydrogels are required so that their properties can be tailored for optimal cellular behaviour [1] and so that they better mimic the properties of natural tissues that are being replaced for tissue engineering applications [5]. The advantages of nanoindentation are numerous including that small volumes of material can be tested with a high spatial resolution, ease of sample preparation, the ability to apply extremely low forces and the ability to map heterogeneous or spatially graded properties of materials [5]. Hence, the technique has significant potential over macroscopic mechanical tests such as rheology, which are typically used to characterise the shear properties of hydrogels. However, the technique, when applied to soft materials and tissues, has a number of limitations because commercial instruments were originally developed for characterising stiff, engineering materials. The main issues faced when characterising compliant materials is accurate surface detection and sample roughness [3].

Recently, dynamic nanoindentation testing has been used to characterise the storage and loss modulus of a range of hydrogel samples which had storage modulus values in the range of 3.5–8.9 MPa [5]. In this paper, we make use of an ultra-low nanoindentation head equipped with a dynamically compliant indenter head operated at its resonant frequency to test the properties of highly compliant gels including low molecular weight hydrogels, which have a shear modulus in the order of tens to hundreds of kPa. The application of this method for potential characterisation of spatially resolved properties of highly compliant gels which serve a range of biomedical applications is demonstrated.

R. Akhtar (✉)
Centre for Materials and Structures, School of Engineering, University of Liverpool, Liverpool L69 3GH, UK
e-mail: r.akhtar@liverpool.ac.uk

E.R. Draper • D.J. Adams
Department of Chemistry, Centre for Materials Discovery, University of Liverpool, Liverpool L69 7ZD, UK

H. Pfaff
Keysight Technologies GmbH, Lyoner Str. 20, Frankfurt 60528, Germany

S.A. Tekalur et al. (eds.), *Mechanics of Biological Systems and Materials, Volume 6*,
Conference Proceedings of the Society for Experimental Mechanics Series, DOI 10.1007/978-3-319-21455-9_17

17.2 Materials and Methods

17.2.1 *Materials*

17.2.1.1 Gelatin Gel

Gelatin (48723-500G-F, Sigma-Aldrich) gel was made with a concentration of 0.34 g/ml. The gel solution was poured into a backless tray and cured in the −20 °C refrigerator. The gel was tested 2 h after removal from the refrigerator.

17.2.1.2 Hydrogels

Two different gelators were utilised in this study (2-naphthalene-valine-glycine (2-NapVG) and 1-tetrahydronathyl-diphenylalanine (1-THNapFF). The gels were made by dissolving 5 mg/mL of the gelator using 1 molar equivalent of NaOH and then making up the solution to 2 mL with deionized water. The solutions were gently shaken until all the gelator was dissolved. Subsequently, 8 mg/mL of glucono-d-lactone (GdL) was added to the solutions to lower the pH and they were left to gel overnight. Nanoindentation tests were carried out on the gels the following day.

17.2.2 *Experiment Setup*

A Nanoindenter G200 system was used with a DCM-II head (Keysight Technologies, USA) for all experiments. A 100 μm flat-ended cylindrical punch tip (Synton-MDP Ltd, Nidau, Switzerland) was mounted on the DCM-II. A custom sample holder was designed into which the gel was transferred. The well in the sample holder was 1 cm^2 and 4 mm deep.

17.2.3 *Nanoindentation Test Procedure and Analysis*

The DCM-II head was used to perform dynamic (oscillatory) indentation and determine the complex shear modulus (G*), which has real and imaginary components, and gives the intrinsic elastic (G′) and viscous (G″) properties of the material:

$$G^* = G' + iG'' \tag{17.1}$$

The theoretical basis for this method has been covered in detailed elsewhere [6] and is summarised here. By indenting samples with a flat-ended cylindrical punch, the relationship between the shear modulus (*G*′), Poisson's ratio (v), elastic contact stiffness (*S*), and punch diameter (*D*) can be determined using Sneddon's analysis [7]:

$$G' = S(1-v)/(2D) \tag{17.2}$$

And a similar relationship for G″ can be derived by considering the contact damping (*Cw*) [8–10]:

$$G'' = Cw(1-v)/(2D) \tag{17.3}$$

The contact stiffness (*S*) is obtained by subtracting the instrument stiffness (*Ki*) from the total measured stiffness (*Ks*):

$$S = Ks - Ki \tag{17.4}$$

The contact damping (*Cw*) is obtained by subtracting the instrument damping (*Ciw*) from the total measured damping (*Csw*):

$$Cw = Csw - Ciw \tag{17.5}$$

Both the stiffness and damping are obtained by oscillating the indenter electromagnetically. Initially, the stiffness and damping of the head only are measured under the same conditions that are subsequently used for testing i.e., the same position, frequency and oscillation amplitude. The indenter head is subsequently moved down towards the sample until contact is sensed. At this point a pre-compression of is applied such that the indenter is in full contact with the sample surface. The

Fig. 17.1 Example profile of phase angle as a function of time that is used for accurate surface detection (S)

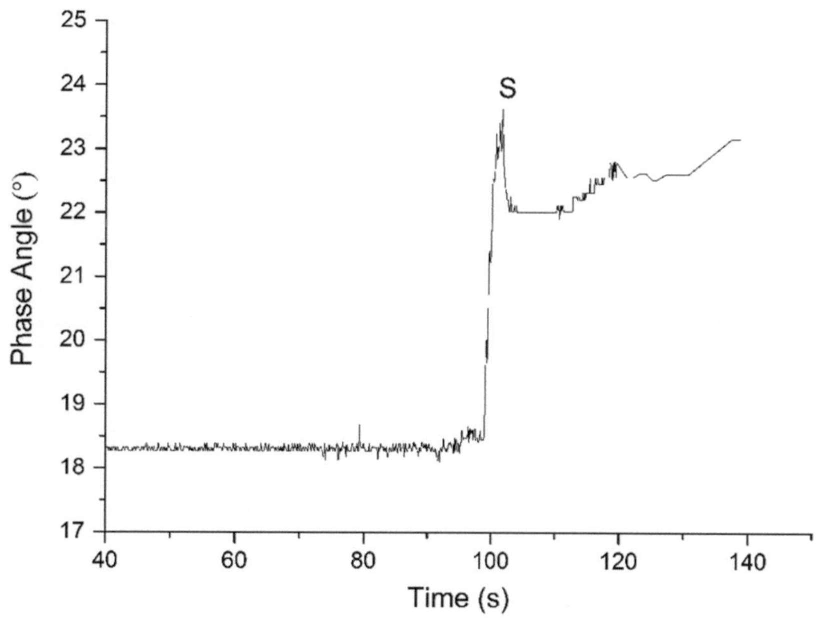

indenter is then vibrated at fixed frequency and oscillation amplitude. Hence, with this two-step process the indenter head is calibrated in situ with the initial oscillation sequence. Stiffness and damping contributions from the head can then be subtracted from the measured signal to obtain the response of the contact only. Subsequently, the second oscillation sequence acquires the material properties.

For the experiments conducted in this study a pre-compression of 5 μm was selected. The indenter was fully in contact with the sample surface. The indenter was then vibrated at a frequency of 110 Hz (the resonant frequency of the indenter) and with an oscillation amplitude of 500 nm. The surface detection relies on a phase shift of the displacement measurement (Fig. 17.1).

As a criterion for detecting the surface, the phase shift was monitored over a number of data points in order to exclude random spikes which were found to occur in some instances over short time intervals (e.g., 2 data points) and thereby lead to an erroneous surface detection. Once the surface detection requirement had been fulfilled over a predefined number of data points the initial contact was determined from the first point in that sequence.

Ten indents were performed on each sample with 200 μm spacing between each sample. A Poisson's ratio of 0.5 was assumed for each of the gel samples. G′, G″ and the loss factor [tan(δ)] i.e., ratio of G″/G′ were calculated for each indentation. After each indent, the tip was cleaned by indenting a piece of double-sided Scotch tape mounted on an adjacent sample puck before returning to the gel sample.

17.3 Results and Discussion

The gelatin gel which serves as a model for soft tissue [11] yielded values for G′ and G″ of 21.7 ± 2.7 kPa and 5.35 ± 0.5 kPa respectively. The loss factor was found to be 0.25 ± 0.01. These values are comparable with those reported in the literature with other techniques [12].

The data obtained with the two different hydrogels is shown in Fig. 17.2.

As shown in Fig. 17.2, the storage modulus (G′) dominates over the loss modulus (G″), which is the expected behaviour of a hydrogel [13]. 2NapVG was found to have a much higher elastic and viscous component than 1THNapFF as well as a slightly higher damping capacity (as determined from the loss factor, Fig. 17.1c). The rheology data for the 2NapVG hydrogel is presented as dipeptide 13 in Table 1 in [13]; G′ = 36 kPa and G″ = 8.1 kPa). The values obtained with nanoindentation are much higher for both G′ and G″ which is likely to be an effect of frequency dependency. The rheology experiments were conducted at 10 rad/s (1.6 Hz) whereas the nanoindentation measurements were performed at 110 Hz, which is the resonant frequency of the indenter. For the 1THNapFF gel, the elastic component of the shear modulus G′ was found to be very com-

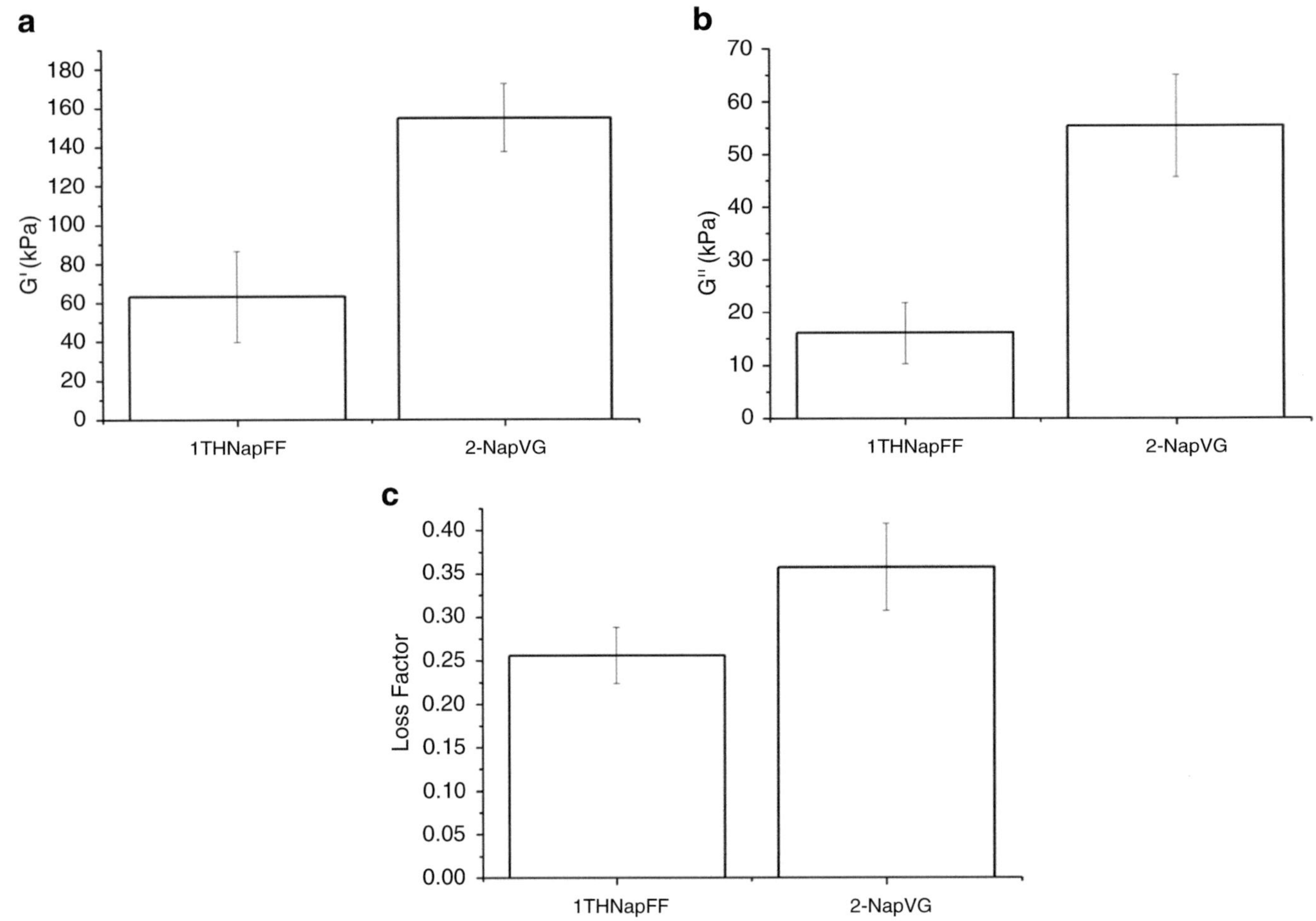

Fig. 17.2 Complex shear modulus data for the two peptide hydrogels (**a**) G′ (**b**) G″ (**c**) Loss factor

parable with rheology measurements that were conducted previously (unpublished data; G′ = 73 kPa also at 10 rad/s using the method detailed in [13]). The value for G″ measured with nanoindentation was much higher than that determined with rheology (2.8 kPa) which is also likely to be related to the much higher frequency at which the nanoindentation tests were performed. Overall the data appear comparable with published data on similar hydrogel systems with conventional rheology [13]. Further, the modifications of the Keysight Technologies standard gel method that were implemented in this study allowed for a more robust way of accurately detecting the sample surface.

Future work will involve characterising the gels across a range of frequencies using both the nanoindentation method presented in this paper and rheology. The methodology may also be useful for soft tissue characterisation.

17.4 Conclusions

In this study, a dynamic nanoindentation method which exploits the unique features of the DCM-II indenter head actuator has been found to be a suitable technique for characterising the elastic and viscous properties of low molecular weight hydrogels and similar materials.

Acknowledgements RA is grateful to the Royal Society for funding (Research Grant. RG130629), and also to the Academy of Medical Sciences for providing travel funding to present this paper at the SEM Annual Conference 2015. Kikelomo Moronkeji must be thanked for preparing the gelatin gel. ERD thanks the EPSRC for a DTA studentship. DJA thanks the EPSRC for a Fellowship (EP/L021978/1).

References

1. Kaufman, J.D., Miller, G.J., Morgan, E.F., Klapperich, C.M.: Time-dependent mechanical characterization of poly (2-hydroxyethyl methacrylate) hydrogels using nanoindentation and unconfined compression. J. Mater. Res. **23**, 1472–1481 (2008)
2. Selby, A., Maldonado-Codina, C., Derby, B.: Influence of specimen thickness on the nanoindentation of hydrogels: measuring the mechanical properties of soft contact lenses. J. Mech. Behav. Biomed. Mater. **35**, 144–156 (2014)
3. Akhtar, R., Schwarzer, N., Sherratt, M.J., Watson, R.E., Graham, H.K., Trafford, A.W., et al.: Nanoindentation of histological specimens: mapping the elastic properties of soft tissues. J. Mater. Res. **24**, 638–646 (2009)
4. Ebenstein, D.M., Pruitt, L.A.: Nanoindentation of soft hydrated materials for application to vascular tissues. J. Biomed. Mater. Res. A **69**, 222–232 (2004)
5. Guglielmi, P.O., Herbert, E.G., Tartivel, L., Behl, M., Lendlein, A., Huber, N., Lilleodden, E.T.: Mechanical characterization of oligo (ethylene glycol)-based hydrogels by dynamic nanoindentation experiments. J. Mech. Behav. Biomed. Mater. **46**, 1–10 (2015)
6. Hay J.: Complex Shear Modulus of Commercial Gelatin by Instrumented Indentation. Agilent Technologies (2011)
7. Sneddon, I.N.: The relation between load and penetration in the axisymmetric boussinesq problem for a punch of arbitrary profile. Int. J. Eng. Sci. **3**, 47–57 (1965)
8. Loubet, J.L., Oliver, W.C., Lucas, B.N.: Measurement of the loss tangent of low-density polyethylene with a nanoindentation technique. J. Mater. Res. **15**, 1195–1198 (2000)
9. Herbert, E.G., Oliver, W.C., Lumsdaine, A., Pharr, G.M.: Measuring the constitutive behavior of viscoelastic solids in the time and frequency domain using flat punch nanoindentation. J. Mater. Res. **24**, 626–637 (2009)
10. Herbert, E.G., Oliver, W.C., Pharr, G.M.: Nanoindentation and the dynamic characterization of viscoelastic solids. J. Phys. D Appl. Phys. **41**, 074021 (2008)
11. Quan, K.M., Christison, G.B., Mackenzie, H.A., Hodgson, P.: Glucose determination by a pulsed photoacoustic technique—an experimental-study using a gelatin-based tissue phantom. Phys. Med. Biol. **38**, 1911–1922 (1993)
12. Markidou, A., Shih, W.Y., Shih, W.H.: Soft-materials elastic and shear moduli measurement using piezoelectric cantilevers. Rev. Sci. Instrum. **76**, 064302 (2005)
13. Houton, K.A., Morris, K.L., Chen, L., Schmidtmann, M., Jones, J.T.A., Serpell, L.C., et al.: On crystal versus fiber formation in dipeptide hydrogelator systems. Langmuir **28**, 9797–9806 (2012)

Chapter 18
Characterization of Acoustically-Induced Forces of the Human Eardrum

Morteza Khaleghi, Cosme Furlong, Jeffrey Tao Cheng, and John J. Rosowski

Abstract Human eardrum or Tympanic Membrane (TM) is a thin structure located at the boundary between outer and middle ears. Shape, deformations, and thickness of the mammalian TMs have been studied by several groups; however, sound-induced forces of the TM, and the question of "how large the forces produced by acoustic waves are along the manubrium at the input to the middle-ear ossicular system?" have not been fully answered. In this paper, sound-induced forces in the human TM are measured at different tonal frequencies and at several points on its surface. A calibrated force sensor with a resolution of 0.5 μN is used with a 3D nano-positioner, enabling accurate placing of the sensor at points of interests on the TM surface. A closed-loop control system is designed and implemented in order to realize constant preload of the sensor at all the measuring points. Concomitant to the force measurements, time-averaged and three-dimensional stroboscopic holographic interferometry are used to compare the modal shape of the sound-induced motion of the TM before and after the presence of the force sensor. The preliminary results show that the maximum sound-induced forces at the umbo occurs at frequencies between 1.5 and 2.3 kHz, whereas the maximum forces for locations on the surface of the TM occurs at around 5–6 kHz.

Keywords Digital holographic interferometry • Human eardrum • Micro-scale force measurements • Sound-induced motions

18.1 Introduction

Mammalian eardrums or Tympanic Membranes (TM) are thin structures (with thickness varying from 20 to 120 μm [1]) located at the boundary between outer and middle ears. Shape, deformations, and thickness of the terrestrial vertebrate TMs have been studied by several groups including our group [2–4]; however, characterization of sound-induced forces in the middle ear have been always challenging and, to the best of our knowledge, there are little or no reports or publications in this field. Some of the challenging issues relating to such force measurements include the dimensions and location of the TM and its viscoelastic properties with relaxation and creep behavior during loading [4]. We have developed a micro-force measuring system to quantify sound-induced forces in the human TM at different tonal frequencies and at several points on its surface. The measuring system uses a MEMS force sensor integrated with a 3D nanopositioner to perform automatic force

M. Khaleghi (✉)
Mechanical Engineering Department, Center for Holographic Studies and Laser micro-mechaTronics (CHSLT), Worcester Polytechnic Institute, Worcester, MA 01609, USA
e-mail: khaleghi29@gmail.com

C. Furlong
Mechanical Engineering Department, Center for Holographic Studies and Laser micro-mechaTronics (CHSLT), Worcester Polytechnic Institute, Worcester, MA 01609, USA

Eaton-Peabody Laboratory, Massachusetts Eye and Ear Infirmary, Boston, MA 02114, USA

Department of Otology and Laryngology, Harvard Medical School, Boston, MA 02114, USA

J.T. Cheng • J.J. Rosowski
Eaton-Peabody Laboratory, Massachusetts Eye and Ear Infirmary, Boston, MA 02114, USA

Department of Otology and Laryngology, Harvard Medical School, Boston, MA 02114, USA

S.A. Tekalur et al. (eds.), *Mechanics of Biological Systems and Materials, Volume 6*,
Conference Proceedings of the Society for Experimental Mechanics Series, DOI 10.1007/978-3-319-21455-9_18

compensation to account for the TM's relaxation by a closed-loop control. Concomitant to the force measurements, time-averaged and three-dimensional stroboscopic holographic interferometry [2] are used to compare the modal vibrational patterns of the sound-induced motion of the TM before and after the presence of the force sensor to monitor how the force sensor is changing the kinematics of the TM motions, and more importantly, to relate the sound-induced displacements data to the measured force data in order to develop constitutive equations. One important outcome of this work is a test of the suggestion that prosthetic replacement eardrums can be improved if they maintain the normal curvature and the material properties of a healthy TM.

18.2 Methods

The experiments include the use of continuous tonal sound stimuli with measurements of the shape and 3D sound-induced motion before and after the use of the force sensor at various locations on the TM surface. The combinations of these force and motion measurements under the different stimulus conditions with measurements of shape and thickness will define: (a) How sound energy is transferred to different locations on the TM [5]; (b) What is the timing relation (possible phase lags) of force and motion in the TM surface; (c) Whether the tent-like shape of the TM acts as a lever which concentrates force at the TM rim and ossicular connections while allowing large motions of the region between the TM rim and its center; and (d) How the thickness and stiffness and viscous moduli of the TM vary over its surface.

18.2.1 Force Measurements

A commercially-available single-axis MEMS-based force sensor (FemtoTools FT-S1000 [6]), shown in Fig. 18.1, with a measuring range of ±1000 μN and a resolution of 0.5 μN is used. The sensor is selected based on both its measuring range and dynamic response that has a flat frequency response from DC up to 10 kHz. Both compressive and tensile forces are measured using a differential capacitive readout mechanism that results in low sensitivity to changes in the environmental conditions. The sensor consists of an actuator suspended by fourfolded springs attached to an outer frame. A force applied to the probe in its axial-direction results in a relative motion of the actuator and a set of capacitive electrodes to transduce motion into capacity changes, which are read out by an analog integrated circuit [7]. Capacitance is a measure of the electrical charge between two conductors separated by an air gap. A load applied to the sensor causes a deflection and as the conductors are moved closer to or farther from one another, the air gap changes, and so does the capacitance. The change in capacitance is converted into an output voltage by the readout electronics. Due to the single-crystalline silicon structure of the sensor, the results are highly repeatable and the sensors are less likely to degrade over time.

18.2.2 Development of the Experimental System

Full-field-of-view, 3D, sound-induced displacements of the TM are measured with the method of multiple illumination directions in holographic interferometry, which have been described in detail in [2]. In order to measure the three components of the displacement vector, $\boldsymbol{d}$, at least three independent measurements with different sensitivity vectors are required. In our approach, and to minimize experimental errors, optical phase maps are obtained with four sensitivity vectors to form an over determined system of equations that is solved with the least-squares error minimization method with

$$\{d\} = [[\mathbb{S}]^T[\mathbb{S}]]^{-1} \times \{[\mathbb{S}]^T\{\Omega\}\} \tag{18.1}$$

Where $[\mathbb{S}]$ is the sensitivity matrix containing all the sensitivity vectors $\boldsymbol{Ki}$, and $\{\Omega\}$ is the fringe-locus function vector.

The sensor has a 3 mm long silicone probe that is positioned on the umbo and other points on the medial and lateral surfaces of the TM preparation. This device is mounted on a 3D positioner that can be remotely controlled along three axes with a resolution of 35 nm and integrated with the holographic system, as shown in Fig. 18.2. Due to the delicate structure of the TM, the approaching and positioning of the force sensor on the TM and the force measurements should be automated in order to avoid any potential damage to the membrane. Furthermore, due to the viscoelastic properties and relaxation of the TM, the preloading of the force sensor has to be such that it remains constant during the measurements while also preventing

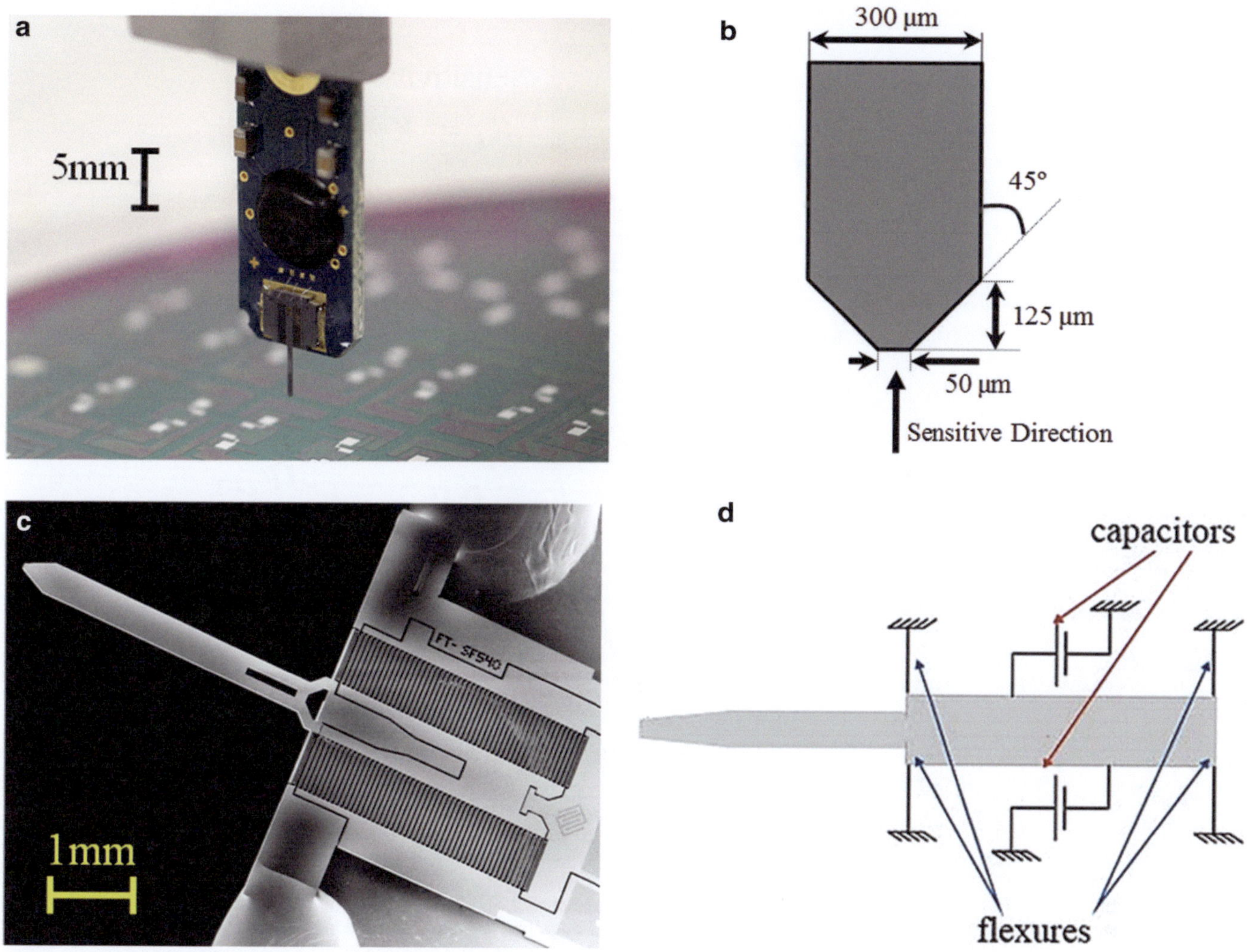

Fig. 18.1 FemtoTools FT-S1000 force sensor used to characterize sound-induced forces in the human TM: (**a**) an image of the force sensor showing characteristic dimension; (**b**) the geometry of the tip of the sensor showing its sensitive direction; (**c**) a microscopic image of the tip of the sensor and its comb structure. The maximum allowable axial deflection of the sensor is 1.5 μm; and (**d**) the internal structure of the sensor. Forces are measured based on capacitive changes induced by motion of the tip [7]

separation from the TM during acoustic loading. Therefore, a closed-loop control is designed and implemented to realize first, automatic approach, and second, constant preloading throughout entire force measurements cycles. As shown in Fig. 18.3, the target preloading, DC_T, and its tolerance, DC_tol, are specified by the user in a developed LabView-based program. The nanopositioner automatically moves the force sensor to maintain a preloading within DC_T ± DC_tol during force-displacements measurements.

18.3 Results

18.3.1 Artificial Membrane

Before application of this force measuring system on real biological samples such as human TM, the sound-induced forces of an artificial sample are measured. The sample is a circular latex membrane with a radius of 5 mm and clamped around its perimeter. While the force sensor is approaching to the membrane, the object is monitored with holographic system in Double-Exposure mode, i.e., the subtraction of the reconstructed hologram from a reference reconstructed hologram.

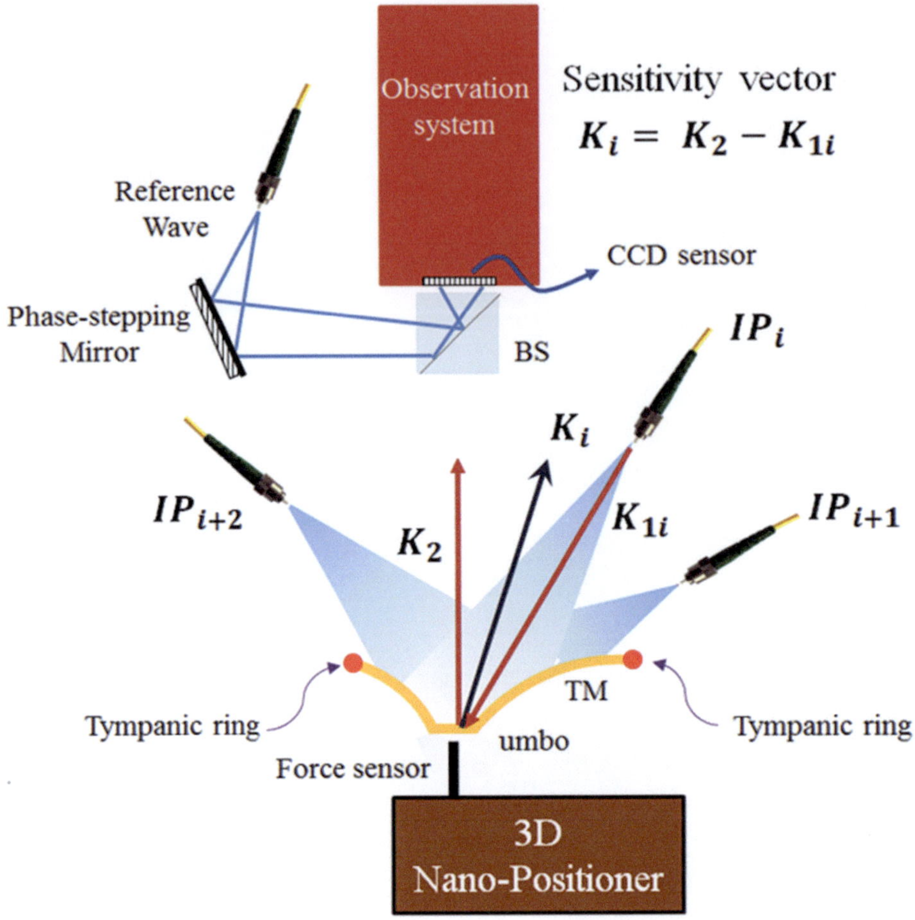

Fig. 18.2 Schematic of the developed holographic system for shape, sound-induced 3D displacement and forces of the human TM. The force sensor is mounted on a 3D nano-positioner with a resolution of 35 nm. The Holographic system is used to both show the position of the force sensor on the membrane and also to gather shape and 3D displacements data before and after the presence of the force sensor

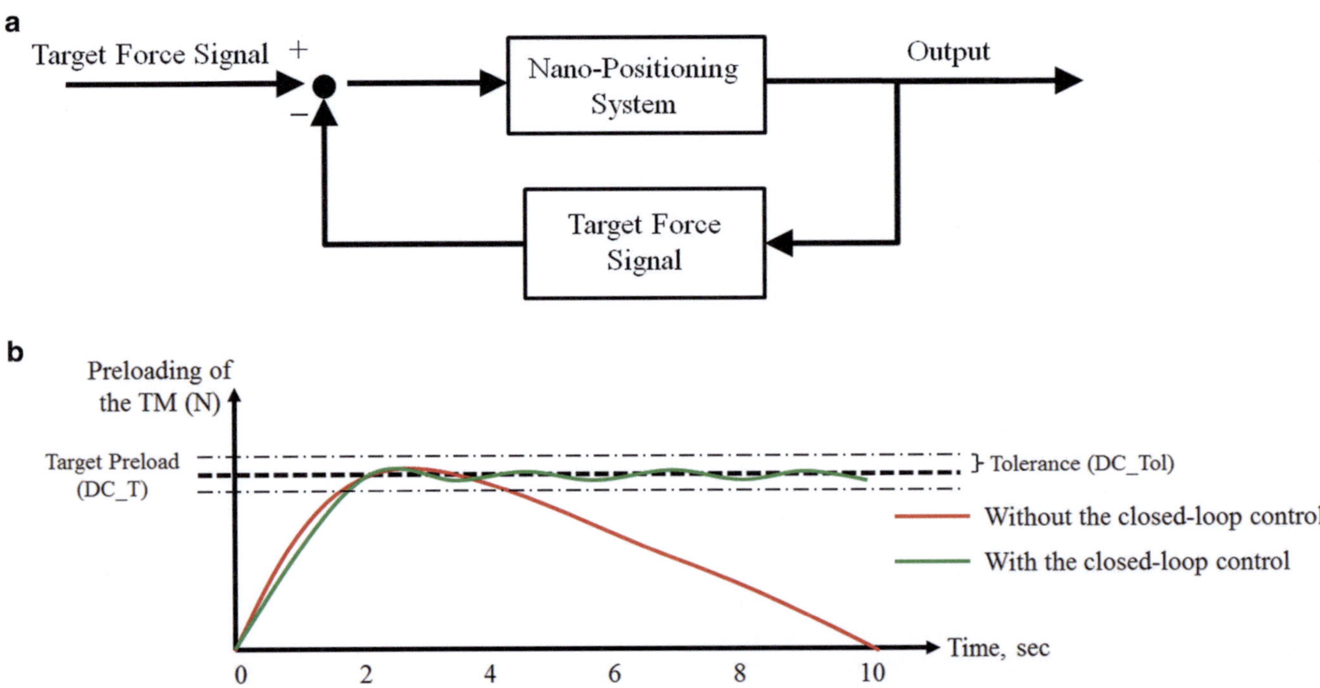

Fig. 18.3 Closed-loop control scheme to compensate for relaxation of the TM during the measurements: (**a**) schematic of the closed-loop control; and (**b**) representative comparisons showing the ability of the developed closed-loop control to maintain constant preloading

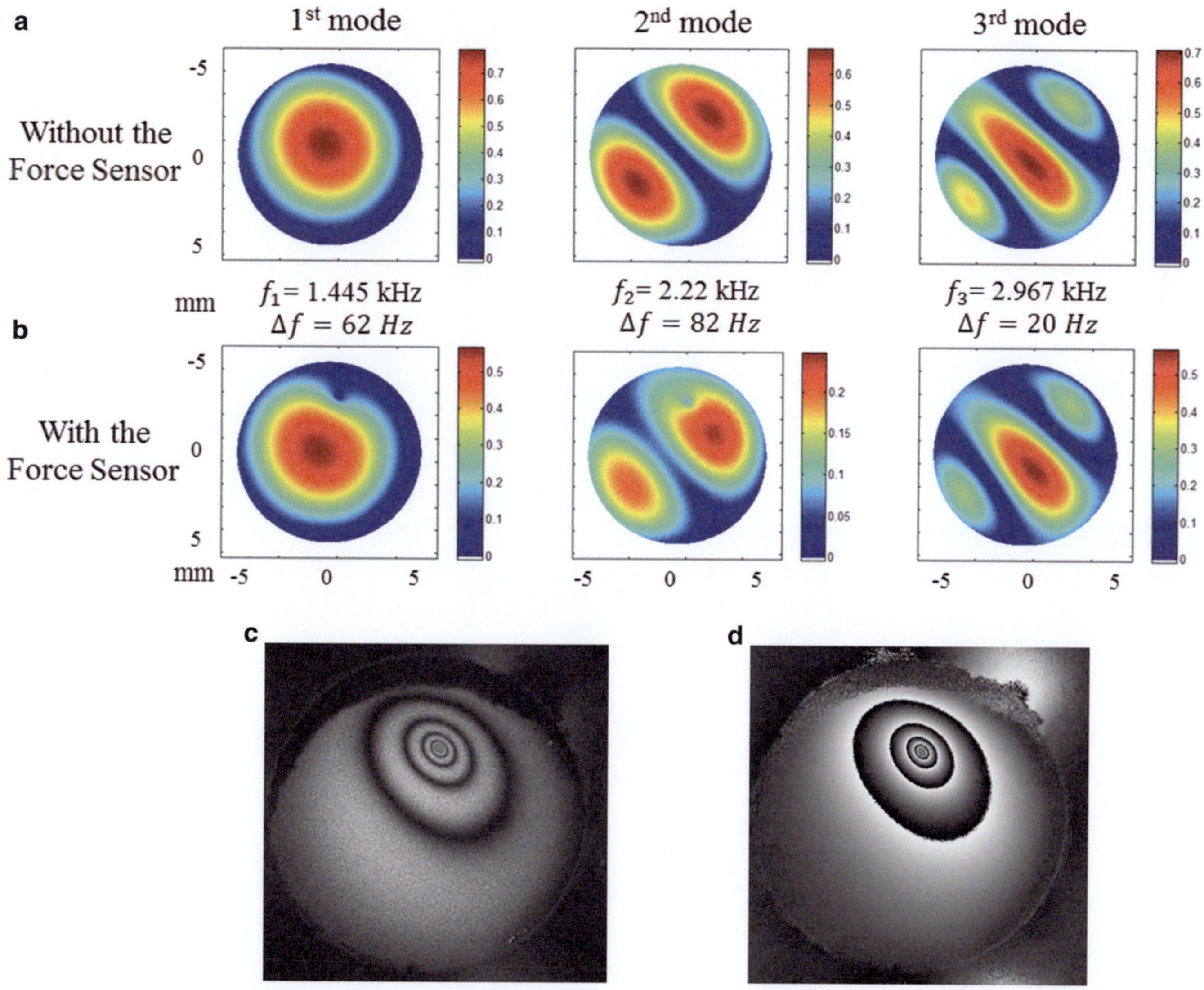

Fig. 18.4 Sound-induced displacement measurements of a circular latex membrane having a diameter of 10 mm and approximate thickness of 20 μm with and without the presence of the force sensor: (**a**) and (**b**) first, second, and third modes of vibration of the membrane before (*top*) and after (*bottom*) the presence of the force sensor, respectively. In (**b**) the location of the sensor is distinguishable. The effects of the force sensor include an increase of the modal frequencies by 1–4 % and a reduction in the amplitude of displacements by approximately 15 % for an applied preload of 250 μN; (**c**) and (**d**) modulation and wrapped optical phase of a reconstructed double exposure hologram corresponding to deformations of the membrane produced by preloading. The applied preload of the membrane is 250 μN producing a displacement of 1.6 μm

As shown in Fig. 18.4, once the tip of the force sensor reaches to the latex membrane, the holographic fringes are produced, indicating the contact of the sensor with the membrane. Then, sound-induced forces and full-field motion of the membrane are measured and the results are compared for both cases of with and without the presence of the force sensor. As shown in Fig. 18.4b the presence of the force sensor increases the apparent stiffness of the membrane, which causes an increase (1–4 %) of its modal frequencies.

18.3.2 Human Tympanic Membrane

A cadaveric human TM of a male, 53 year-old donor was prepared by removing the bony portion of the ear canal. As shown in Fig. 18.5, the incus and stapes of the middle ear ossicular chain have been also removed in order to enable positioning of the force sensor on different points on the medial side of the TM. Due to semi-transparency of the mammalian TMs, the

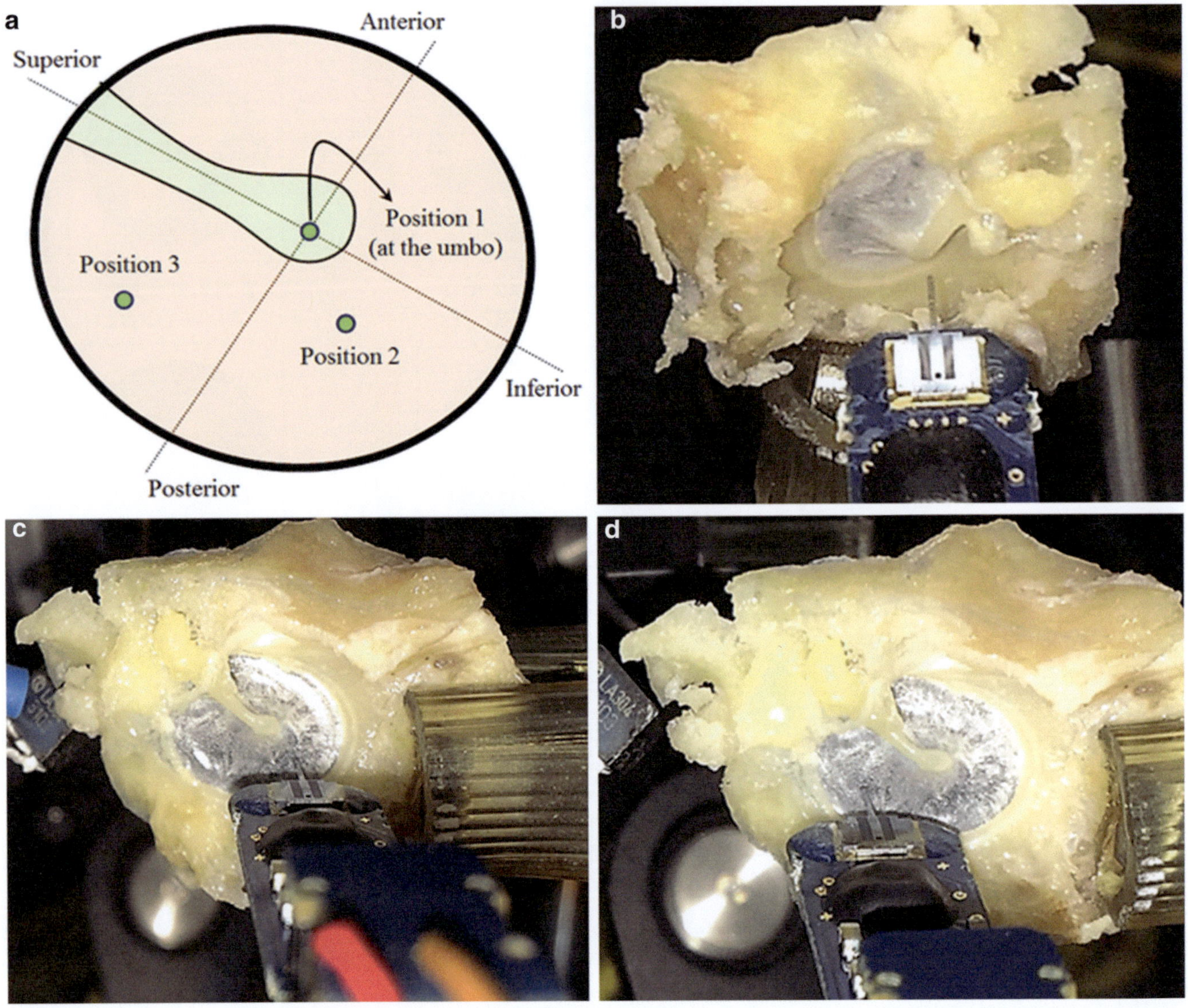

Fig. 18.5 The force sensor is located at different locations of the TM: (**a**) schematic of TM showing the locations of all three target points; (**b**)–(**d**) shows, respectively, the images corresponding to automatic positioning of the force sensor on position 1 (at the umbo), position 2 (in the inferior-posterior) and position 3 (in the posterior-superior)

sample was coated with a thin layer of zinc oxide to increase the laser light reflection. The effect of coating on shape and deformation patterns have been studied by several researchers and found to be negligible [8–10]. The temporal bone was held with an adjustable clamp and mounted on a post in front of the holographic system. As was described in Section 18.2, the force sensor is mounted on the 3D nano-positioner, which is remotely controlled with a computer to place the sensor at different points of interest on the TM.

Prior to stroboscopic measurements, time-averaged response of the TM at different tonal stimuli was monitored [3, 11] and the excitation frequencies were chosen at the maximum motion of the TM. At each stroboscopic phase, sound-induced motions of the TM along three orthogonal axes x, y and z are calculated from the unwrapped optical phases. Then, a numerical rotation matrix is used to calculate the motion components tangent and normal to the local plane of the TM [2, 12]. Definitions of in-plane and out-of-plane motion components have been described in detail in [2]. In-plane components are the ones tangent to the local plane of the TM, whereas the out-of-plane components are normal to the local plane of the TM (i.e., along the local normal vector to the shape of the TM). As shown in Fig. 18.6, the presence of the force sensor reduces the maximum amplitude of vibration of the TM by less than 10 %. This phenomenon can be explained

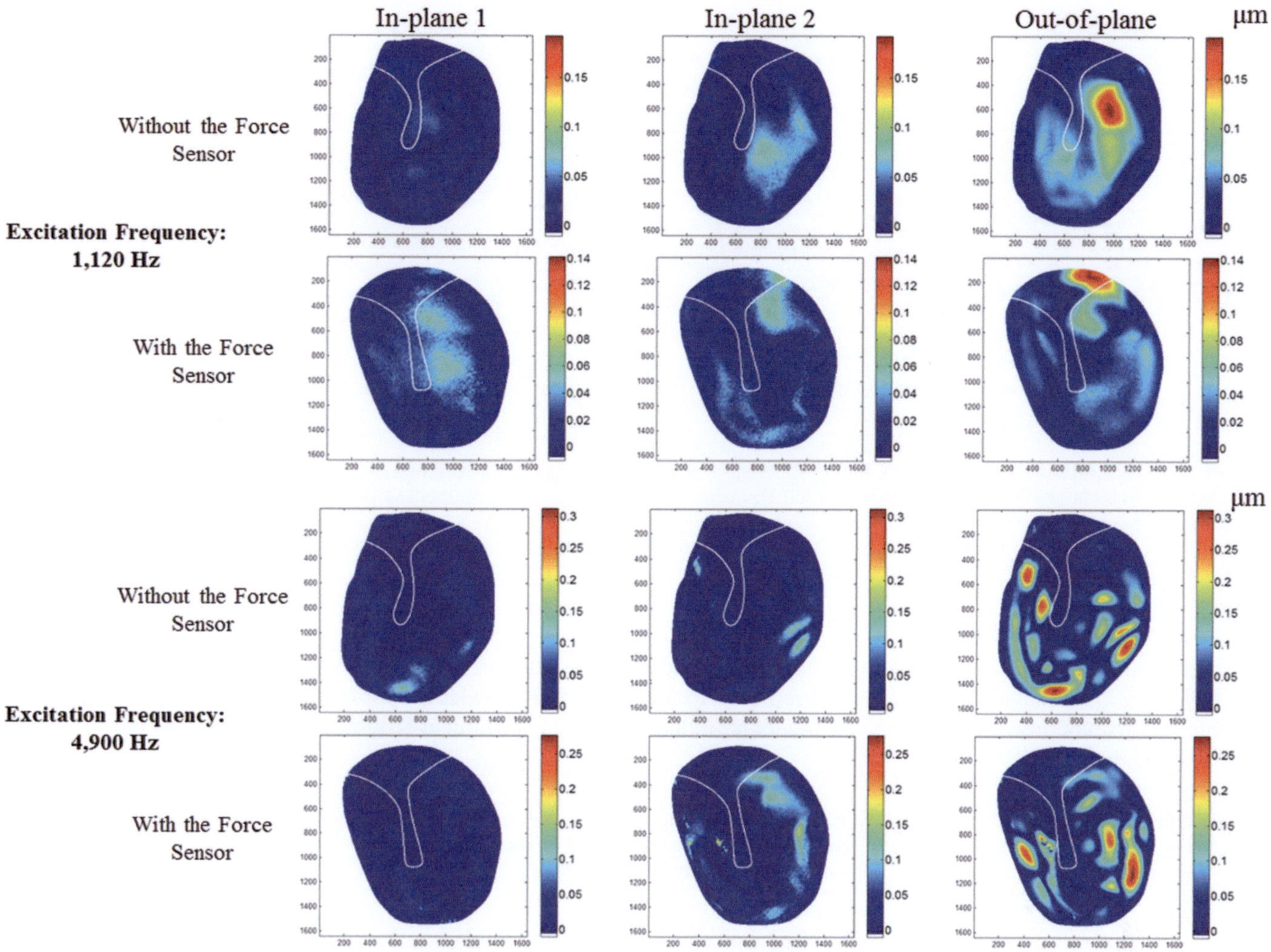

Fig. 18.6 Sound-induced in-plane and out-of-plane motions of a human TM for two different frequencies before and after the presence of the force sensor as measured with stroboscopic holographic interferometry. In-plane 1 and 2 are displacement components tangent, and out-of-plane, perpendicular to the plane of the TM. The presence of the force sensor at the umbo reduces the amplitude of vibration by less of 10 % and changes in the deformation patterns. Magnitudes of displacements are less than 300 nm produced by a sound pressure of 105 dB SPL

by the fact that the presence of the force sensor increases the overall stiffness of the system, which in turn, decreases the amplitude of vibration.

Figure 18.7 shows measured sound-induced forces generated at three different locations on a human TM surface at different tonal frequencies from 300 Hz to 10 kHz. As shown in Fig. 18.7, the frequency-dependent measured forces are also spatially-dependent. For instance, the sound-induced forces at the umbo (position 1—shown with circular markers) have a maximum value at around 1.8 kHz, whereas sound-induced forces at points on the surface of the eardrum (positions 2 and 3) have their maximum values at higher frequency ranges, i.e., 6.2 and 4.7 kHz for positions 2 and 3, respectively.

18.4 Conclusions

We have developed a micro-force measuring system capable of characterizing sound-induced forces in the human TM at different tonal frequencies and at several points on its surface. A close-loop control system is developed to compensate the relaxation and creep behavior of the TM and in order to realize a constant amount of preload in all the measuring points. The force measuring capabilities are combined with digital holographic measurements of shape and true 3D displacements to help expand our knowledge of the hearing processes. The results show that frequency-dependent forces in the middle ear,

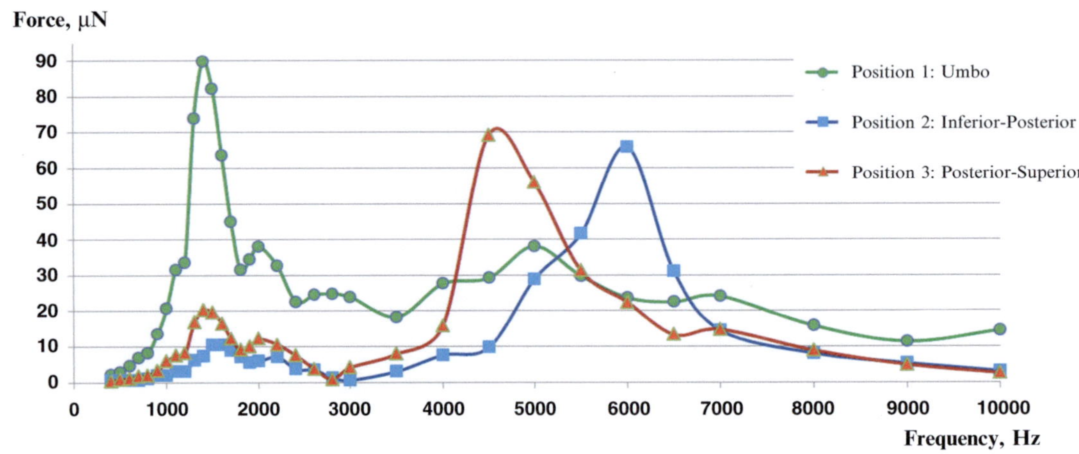

Fig. 18.7 Normalized sound-induced forces at three different locations on a human TM at different excitation frequencies. The maximum force value for the umbo is at excitation frequency of 1800 Hz, whereas forces at locations on the TM surface have peaks at higher frequencies. Measuring repeatability is within ±3 %

and in particular the TM, are also spatially dependent. The effects of the force sensor on the dynamic response of the TM will be further investigated as part of our future work. To the best of our knowledge, our force measurements are one of the first attempts to experimentally characterize the sound-induced forces on the human TM toward understanding their coupling into the ossicular chain.

Acknowledgments This work was supported by the National Institute on Deafness and other Communication Disorders (NIDCD), Massachusetts Eye and Ear Infirmary (MEEI), and the Mechanical Engineering Department at Worcester Polytechnic Institute. We also acknowledge the support of all of the members of the CHSLT labs at WPI and Eaton-Peabody labs at MEEI.

References

1. Van der Jeught, S., Dirckx, J.J., Aerts, J.R., Bradu, A., Podoleanu, A.G., Buytaert, J.A.: Full-field thickness distribution of human tympanic membrane obtained with optical coherence tomography. JARO **14**(4), 483–494 (2013)
2. Khaleghi, M., Furlong, C., Ravicz, M., Cheng, J.T., Rosowski, J.J.: Three-dimensional vibrometry of the human eardrum with stroboscopic lensless digital holography. J. Biomed. Opt. **20**(5):051028 (2015)
3. Rosowski, J.J., Dobrev, I., Khaleghi, M., Lu, W., Cheng, J.T., Harrington, E., Furlong, C.: Measurements of three-dimensional shape and sound-induced motion of the chinchilla tympanic membrane. Hear. Res. **301**, 44–52 (2013)
4. Aernouts, J., Aerts, J.R., Dirckx, J.J.: Mechanical properties of human tympanic membrane in the quasi-static regime from in situ point indentation measurements. Hear. Res. **290**(1), 45–54 (2012)
5. Khaleghi, M., Dobrev, I., Harrington, E., Furlong, C., Rosowski, J.J.: Study of the transient response of Tympanic Membranes under acoustic excitation. In: Barthelat, F., Zavattieri, P., Korach, C.S., Prorok, B.C., Jane Grande-Allen, K. (eds.) Mechanics of Biological Systems and Materials, vol. 4, pp. 1–9. Springer, Cham (2014)
6. FemtoTools Website.: http://www.femtotools.com/products/accessories/ft-s-microforce-sensing-probe. Accessed 10 Mar 2015.
7. Beyeler, F., Muntwyler, S., Dietze, W.: Sub-millinewton capacitive MEMS force sensor for mechanical testing on a microscope. U.S. Patent Application 14/347,421, 2012
8. Dirckx, J.J.J., Decraemer, W.F.: Coating techniques in optical interferometric metrology. Appl. Opt. **36**(13), 2776–2782 (1997)
9. Cheng, J.T., Hamade, M., Merchant, S.N., Rosowski, J.J., Harrington, E., Furlong, C.: Wave motion on the surface of the human tympanic membrane: holographic measurement and modeling analysis. JASA **133**(2), 918–937 (2013)
10. Khaleghi, M., Lu, W., Dobrev, I., Cheng, J.T., Furlong, C., Rosowski, J.J.: Digital holographic measurements of shape and three-dimensional sound-induced displacements of Tympanic Membrane. Opt. Eng. **52**(10), 101916 (2013)
11. Khaleghi, M., Guignard, J., Furlong, C., Rosowski, J.J.: Simultaneous full-field 3D vibrometry of the human eardrum using spatial-bandwidth multiplexed holography. J. Biomed. Opt. **20**(11), 111202 (2015)
12. Alamdari, A., Krovi, V.: Active reconfiguration for performance enhancement in articulated wheeled vehicles. Proc. ASME 2014, Dynamic Systems and Control Conference, V002T27A004-V002T27A004 (2014)

MIX
Papier aus verantwortungsvollen Quellen
Paper from responsible sources
FSC® C105338

If you have any concerns about our products,
you can contact us on
ProductSafety@springernature.com

In case Publisher is established outside the EU,
the EU authorized representative is:
Springer Nature Customer Service Center GmbH
Europaplatz 3, 69115 Heidelberg, Germany

Printed by Libri Plureos GmbH
in Hamburg, Germany